I0152344

REPORT

ON

FOREIGN MANŒUVRES

IN

1912.

GENERAL STAFF, WAR OFFICE

The Naval & Military Press Ltd

Published by the

The Naval & Military Press

in association with the Royal Armouries

Unit 10 Ridgewood Industrial Park,
Uckfield, East Sussex, TN22 5QE
Tel: +44 (0) 1825 749494
Fax: +44 (0) 1825 765701

MILITARY HISTORY AT YOUR FINGERTIPS
www.naval-military-press.com

ONLINE GENEALOGY RESEARCH
www.military-genealogy.com

ONLINE MILITARY CARTOGRAPHY
www.militarymaproom.com

ROYAL
ARMOURIES

The Library & Archives Department at the Royal Armouries Museum, Leeds, specialises in the history and development of armour and weapons from earliest times to the present day. Material relating to the development of artillery and modern fortifications is held at the Royal Armouries Museum, Fort Nelson.

For further information contact:
Royal Armouries Museum, Library, Armouries Drive,
Leeds, West Yorkshire LS10 1LT
Royal Armouries, Library, Fort Nelson, Down End Road, Fareham PO17 6AN

Or visit the Museum's website at
www.armouries.org.uk

CONTENTS.

A.—Direction of Manœuvres.
B.—Staff ; tactics of the three arms combined.
C.—Infantry.
D.—Cavalry.
E.—Artillery.

F.—Engineers ; signal service.
G.—Air service.
H.—Supply, transport, quarters.
I.—Medical.
J.—General impressions.

Notes.

1. In the case of those countries where more than one set of manœuvres was witnessed, descriptions of the various operations are given immediately before A. The sections dealing with infantry, cavalry, &c., contain a summary of the various reports and are not repeated for each set of manœuvres.

2. The contents of each section are arranged, as far as possible, in the following order :—Organization (general) ; tactics (or method of employment) ; personnel ; arms ; equipment ; transport ; inter-communication.

3. Remarks upon machine guns and cyclists are given at the end of the section dealing with that arm (infantry or cavalry) to which they may happen to belong.

INTRODUCTORY NOTE.

This work has been made confidential for the following reasons : In certain cases, officers who furnished reports were given, by the foreign governments concerned, special facilities for acquiring part of the information embodied in this book. It is, therefore, necessary to take special precautions that the book does not fall into the hands of persons who are not officially entitled to peruse it, a contingency which might result in such valuable sources of information being closed in future years.

Copies of the book, therefore, will only be issued to certain officials or deposited in reference libraries, *vide* King's Regulations, paragraphs 1858, 1863. As this will to some extent restrict the circulation of the book, the attention of General Officers Commanding-in-Chief is directed to the necessity of bringing it to the notice of all those under their command who are likely to profit by its perusal.

In the Report on Foreign Manœuvres in 1911, a new system of arrangement was introduced. In this report the system in use previous to 1911 has been followed, but with certain modifications, which represent an attempt to combine the advantages of both systems.

Reports indicate that the opportunities for observation afforded to officers attending foreign manœuvres are becoming more and more limited. Moreover, in some countries, the transport, medical and other services do not take part in manœuvres under war conditions but carry out field exercises earlier in the year. In several cases, again, army manœuvres are designed for spectacular effect, and even in France where they appear to be most realistic, many details of procedure such as ranging by artillery and use of cover by infantry are not carried out.

For these reasons, and in order to give as far as possible a complete impression of recent developments and tendencies in foreign armies, extracts from new regulations, and from the reports of officers attached to foreign units during 1912, are included in this report.

Direction of manœuvres.—Japan and, to a lesser extent, Germany, furnish examples of the system under which manœuvres are carried out according to a carefully pre-arranged programme. The Japanese views on this point are given on page 67. In France, on the other hand, considerable latitude is given to commanders at the army manœuvres (*see* page 10).

Umpires.—The general tendency appears to be to limit the powers and duties of umpires. In Russia, however, elaborate instructions are issued for their guidance (*see* page 99).

In France, umpires were not allotted definitely to formations, but daily to different zones. Their decisions were thus less liable to the charge of bias (*see* page 10).

Maps.—The cheapness of maps available for troops taking part in foreign manœuvres is striking. The maximum price of a map mentioned in any report is 2¾d.

Staff.—The organization and duties of the staff of higher formations in the French Army are given on pages 11–15.

It was noticed that reserve officers, who perform the same duties from year to year, were employed on the staffs of French cavalry divisions.

Reports praise the system of arm-bands for French staff officers, by means of which the headquarters to which an officer belongs can be told at a glance.

Infantry.—The neglect of infantry covering fire in the attack is noticed in nearly every report.

Attention is drawn to the fact that in the German, Japanese and Russian armies the defenders of a position always advance to meet the attacking infantry when the latter are from 30–150 yards distant from the position. This practice has an important bearing upon the question of artillery support during the final stages of the infantry attack

A Russian attack is described on page 101 ; and an account of forest fighting during the Norwegian manœuvres is given on page 92.

Machine guns (infantry).—In Russia, it seems that the value of machine guns is the lesson of the Manchurian war that has made most impression, and not only in Russia but in Austria-Hungary, Germany, Greece, Italy, and Sweden the number of machine guns lately has been or is about to be increased.

The present (1913) proportion of machine guns to 1,000 rifles in different armies is as follows (the letters (C) or (P) indicate whether the equipment is carried on the march on a travelling carriage or on pack animals) :—

Austria-Hungary (P), 2.
France (C), 2.
Germany (C), 2.
Italy (P), 1·25.
Japan (P), 2·5.
Russia (C and P), 4·5.

Machine gun sections mounted on cycles or motor cycles exist in Denmark, France (experimental), and Italy.

A description of the new Russian machine gun organization and equipment is given on pages 104–106.

Cyclists.—The reports show that cyclists are used for fighting and not merely for inter-communication purposes in the following countries : Austria-Hungary, France, Italy, Norway, Sweden (1 company only).

The cyclist units which will be integral parts of French cavalry divisions are described on pages 25–26.

Cavalry.—France, Japan, Norway and Russia have adopted or are about to adopt the bayonet as part of the equipment of the cavalry.

In France all heavy cavalry now carry the lance and the arming of hussars and chasseurs with lances has already begun.

In Russia the front ranks of all regiments will be armed with lances by the spring of 1913.

New regulations for cavalry appeared during 1912 in Russia (*see* page 106) and Japan.

Machine guns (cavalry).—The proportion of machine guns to squadrons is at present (1913) as follows:—

France (C)	..	..	1 gun to 4 squadrons.
Austria-Hungary (P)	..	2	,, 3	,,
Japan (P)	..	.. 1	,, 1	,,
Norway (C)	..	.. 8	,, 13	,,
Russia ..	.	.. 1	,, 3	,,
Germany (C) ..	..	1	,, 4	,,
Italy (P)	..	.. 2	,, 5	,,

Engineers.—An account of bridging exercises in Japan is given on pages 77–80.

Signal service.—Practically all the reports comment upon the neglect of visual signalling.

The Russian method of laying out telephones is described on page 110.

Air service.—The number of aircraft present at the different army manœuvres was as follows:—

Austria-Hungary	..	19 aeroplanes.
France	..	.. 60	,,	2 dirigibles.
Germany	..	.. 38	,,	2 ,,
Greece	..	.. 2	,,
Japan	..	.. 2	,,	1 dirigible.
Norway	,.	.. 1	,,
Russia	..	.. 8	,,	1 ,,
Sweden	..	.. 2	,,
United States	..	3	,,

The organization and work of the French aeroplane service is given on pages 28–33.

The chief lessons gained at the United States manœuvres as regards aeroplanes are given on page 130.

Supply and transport.—A detailed description of the French system of supply and transport is given on pages 33–45.

In France it is proposed to do away with travelling kitchens for the reasons given on page 39. These vehicles were, however, used at the manœuvres in Austria-Hungary, Denmark (grocery wagons), Germany, Norway and the United States, and apparently they gave satisfaction.

Meat wagons were in use at the Danish, French, Norwegian, Swedish and Swiss manœuvres.

Medical.—The method of work and recent changes in the organization of Japanese medical units are given on pages 67, 85–88.

15th February, 1913.

AUSTRIA-HUNGARY.

The grand manoeuvres this year were held in south-east Hungary between the 9th and 11th September, under the direction of the Heir-Presumptive, the Archduke Franz Ferdinand, who was assisted by the Chief of the General Staff, Field-Marshal Blasius Schemua.

The strength and composition of the opposing armies were—

Southern Army (Red).—Commander—General of Infantry Von Koevess. 36 battalions of infantry, 30 machine gun detachments, 30 squadrons, 5 cavalry pioneer sections, 1 cavalry machine gun detachment, 1 cyclist company, 14 field batteries, 2 field howitzer batteries, 3 pioneer companies, 4 heavy pontoon trains, 1 cavalry pontoon train (transported in motor lorries), 2 wireless stations, and 1 aeroplane park of 7 Etrich monoplanes.

On the 9th September, the Red army received reinforcements from the (imaginary) main army, consisting of 2 infantry battalions, $\frac{1}{2}$ squadron, 1 field battery, $\frac{1}{2}$ company of pioneers, and a Danube-Theiss flotilla of 2 river monitors and 2 patrol boats.

Northern Army (Blue).—Commander—Army Inspector-General of Infantry Schoedler. 53 battalions of infantry, 48 machine gun detachments, 28 squadrons, 5 cavalry pioneer sections, 1 cavalry machine gun detachment, 1 cyclist company, 3 horse batteries, 15 field batteries, 4 field howitzer batteries, 4 pioneer companies, 1 heavy pontoon train, 1 cavalry pontoon train, 4 wireless stations, and 1 aeroplane park with 7 Etrich monoplanes.

At the suggestion of the Archduke Franz Ferdinand, this year's grand manoeuvres differed from all previous grand manoeuvres in Austria-Hungary, in that they were held, not in rolling or mountainous country, but in the absolutely flat lowland of southern Hungary, called "Niedungarische Tiefebene" or "Alfoeld," which is said to resemble the plains of upper Italy. At the time of the manoeuvres, the ground was covered with mealie crops from 10 to 12 feet high, a circumstance which had a pronounced effect upon the operations, and prevented onlookers from forming any reliable opinion upon the method of conducting them. The roads were in some cases indicated by avenues of trees, and the position of villages by church spires; otherwise there was little to break the dead monotony of the plain, and the location of objectives was difficult in consequence.

Umpires.—The instructions to umpires were the same as those issued last year. Umpires were not allowed in any way to influence the movements of troops, and all their decisions had to be given in writing. At least one umpire was assigned to each infantry battalion, every two squadrons of cavalry, and each artillery division. During the grand manoeuvres of 1911, some commanders did not accept decisions of umpires who were junior to them in rank. Accordingly, this year umpires were, if possible, of the same rank as the commanders of the units to which they were attached, while 26 general officers were attached to the umpire staff.

The only restrictions upon movement were that troops must have at least 4 hours' rest in every 24 hours. Night operations were permitted.

Infantry and artillery.—As has been already indicated, the close character of the country prevented observation of infantry or artillery fire tactics. The general impression gathered was that, apart from marching power, the infantry lacked many qualities required by troops in modern war. They seemed slow and heavy, and the low proportion of officers was very noticeable. The scarcity of artillery and the entire lack of co-operation between field guns and howitzers were other marked features. Fire was generally directed from observation ladders, which were concealed by sheaves of mealie stalks.

Bridging.—The pontoons did excellent work, the personnel being picked men; the new steel pontoon earned high praise.

Air service.—The work of the aeroplanes attracted particular attention, and was much praised. All the machines used (with the exception of one Lohner "Arrow" biplane) were Etrich monoplanes, and it seems likely that the Austrians will adhere to this type, especially in view of the influence of the inventor, Herr Etrich, and his connection with the Daimler firm, which supplies the engines.

The weather up to the last day was very unfavourable for flying, strong wind and continuous rain prevailing. One pilot took 2 hours to cover a distance of 50 kilometres against the wind, but came back in 15 minutes. Each machine carried a pilot and one specially trained observer. Much information was procured for commanders and was in the main correct. Generally speaking, the bulk of this information was such as could not have been procured by ordinary means of reconnaissance. Aircraft passing over friendly troops dropped reports contained in metal cases, which were provided with a smoke-producing compound, and also a streamer 7 metres long. No difficulty was experienced in making landings or rising from the wet fields. At night, searchlights were employed to illuminate landing places. It is remarkable that during 3 days' continuous flying no accident of any sort occurred. The Etrich monoplane is exceedingly steady during flight, and has given the greatest satisfaction in the Austro-Hungarian service. It is, however, heavy, difficult to transport (because the wings do not fold), and the motor is exceedingly noisy.

Transport.—On account of the poor roads, heavy mechanical transport was very little used, one division only being supplied by this means. The travelling kitchens gave satisfaction, and kept up with units without difficulty.

The Volunteer Automobile Corps did excellent work in carrying despatches, furnishing patrols, &c. They have already (1912) placed at the disposal of the Government no less than 550 motor cars and 250 motor cycles. The service is very popular, as it affords a pleasant manner of finishing the general service obligation.

DENMARK.

The manœuvres took place on Zealand between the 27th September and 4th October. Foreign representatives were invited for 3 days ; the last 4 days included fortress manœuvres of a secret nature.

Practically the sole interest of the 3 days' exercises seen, lay in the facts that they were the first held under the new King, who is above all things a soldier, and that the dispositions of troops give some clue to the strategy that might be adopted to defend the capital in case of a sudden invasion from the north.

The opposing forces were organized as follows :—

Blue (invaders). Commander—Major-General Schroll.

1st and 2nd (Zealand) Divisions—

12 battalions,
4 squadrons,
8 batteries (32 guns),
2 telegraph detachments, and a
Volunteer machine gun detachment, consisting of 20 guns (Rekylgevaer) mounted on motor cycles.

Red (defenders). Commander—Major-General Wolff.

3rd (Zealand) Division (Composite)—

14 battalions.
4 squadrons.
8 batteries (4 marked).
1 engineer company.

Effectives were weak, only two classes (those of 1909 and 1910) being called up for repetition exercises. Companies seldom exceeded 100 men (war strength is 250), and were often less. Squadrons averaged 110 men (war strength, 150).

The country is open and undulating, cultivated and generally unfenced, with small farms and hamlets at short intervals. Roads are numerous and good. It would be difficult to find a country better adapted for manœuvres and the training of troops, abounding as it does in varied topographical features, affording many opportunities for attack and defence.

The general idea was as follows :—

A Blue invader with command of the sea had made a sudden descent on Denmark with Copenhagen as his objective. He had effected a surprise landing on the north coast of Zealand with a detachment of 10 to 12 battalions, four squadrons and eight batteries. Meanwhile, the main Blue army (imaginary) sailed up the Oresund (between Zealand and Sweden) with a view to effecting a landing near the capital.

B 2

The task of the Blue detachment was to engage a Red force known to be about Roskilde (20 miles west of Copenhagen), and to prevent it moving east to interfere with the landing of the main Blue army. The task of the Red force was to prevent the Blue detachment from advancing on the capital.

The main Red army was supposed to be mobilizing, and no support was to be expected from it for the first few days.

Direction of Manœuvres.

From the time operations started, commanders had a free hand to move over the whole of Zealand and to billet where they liked at a moment's notice. Beyond the imaginary features of the scheme no influence apparently was exercised on the course of operations by the directing staff.

Maps.—Ten maps per company were issued free, and many men purchased copies themselves at a cost of 10 *öre* (1¼*d.*). The " war map " that is used by officers, and to which all references are made, is on a scale of 1 : 40,000.

Infantry.

The men are intelligent, sturdy, good-tempered, clean, and look after their equipment and arms well. Everything is done for their comfort and they are perhaps rather spoilt. They are obviously insufficiently trained, and the shortage of officers and under-officers emphasizes this fact when any offensive movement has to be undertaken. The pace of march is slow, but march discipline is good.

Nothing new can be said as regards the infantry tactics except that a form of advance (called " Japanese ") to the final fire position by the dribbling up of single men was practised. Bayonets were fixed when within 600 yards of the enemy; volley firing was heard, and troops were sometimes seen to move to a flank in close formation under effective fire of infantry or artillery.

The barrels of the rifles (1889 pattern adapted for the pointed bullet) were much worn, and complaints were heard as to bad shooting in consequence. It is said that the steel used is too soft. These points have been ventilated in the press, and the Defence Minister has asked for a grant of 250,000 *kroner* (£14,000) for the purpose of replacing worn-out weapons.

Cavalry.

The few squadrons seen were well mounted on English and Swedish horses in excellent condition. Their tactics were somewhat antiquated—patrols invariably moving with drawn swords. Fire action was little practised except to support machine guns. The cavalry has excellent material in men and horses, and a good spirit.

Artillery.

Batteries took the field with only four wagons to save the expense of hiring horses. The Goerz panoramic sight has now been adopted, and all fire was indirect, either from covered or semi-covered positions. The Ehrhardt gun, though now 10 years old, was well spoken of, and detachments appeared to be familiar with it. Observation stations were connected by telephone (three stations with each battery) with batteries. A favourite post of observation was the village church. An ammunition park for the Red Force was established at Roskilde, and supply from it practised for the first time.

Supply and Transport.

Two days' supplies at a time were forwarded by road from the base of the Red Force at Roskilde. Blue supplies were neutral and sent from Copenhagen by rail. The only novelties seen were covered meat wagons in which the carcases were hung, and four experimental mess wagons, containing canteens with eating utensils, for officers and under-officers, and a 100-litre water tank. The grocery wagons reported on last year have been adopted.

Miscellaneous.

Volunteer Corps.—The voluntary organizations are deserving of mention. Nearly every town in Denmark has now a *Frivillige Korps*, and several *Korps* are armed with machine guns on motor cycles. The Westenholz, the model of these, called after the patriotic gentleman who raised and maintains it, consists of 50 members all mounted on motor cycles and armed with the "Rekylgevaer." This gun is carried upright between the handle-bars, while a tray behind the saddle takes the ammunition—500 rounds, in clips; the gun is also used by the cavalry. The Danes only claim that it is light and shoots well up to 1,200 m.; but it stands only some 9 inches from the ground and the vibration is very great. Other countries (*e.g.*, Russia and Belgium) have given it a trial and condemned it.

Two other corps were represented at manœuvres, the Academy Rifles (from the University) who turned out about 80 strong, and the Motor Orderly Corps. There were 25 of the latter who played the same part as our own motor volunteers on motor cycles. These corps must all do 10 days' continuous exercises besides a certain number of drills.

The volunteer movement is recognised by the War Office, and regular officers and under-officers are appointed to the corps as instructors.

Summer Camps.

The new permanent camps in which the summer training of recruits will take place are (for a poor country ill-provided with the technical needs of war) luxurious; they consist of a number of brick and stone bungalows. Ten men sleep in each room, and there are hot and cold baths with "showers." All the cooking and messing arrangements are excellent. A good deal of space is wasted. Camp accommodation for five battalions—2,500 men—costs the country £200,000.

Officers.

Since the large manœuvres of 1908 the Danish Army shows no apparent advance in efficiency, and a good deal of its backwardness, in comparison with other small armies, must be attributed to the officer, who is often indolent and ease-loving. The ignorance of the average officer as to the progress of the science of war in other armies is lamentable. For this state of things the democratic and socialistic condition of the country, its attitude of indifference towards defence, and the enervating effect of contiguity to Germany, are largely responsible. The officers labour under great disadvantages, mostly financial, and those who can afford to go to other countries to bring themselves up to date are few. It is hoped and thought that the King, who has the welfare of the army at heart, will do something to remedy this.

Under-officers.

The under-officers are allowed to stay on up to 50 years of age. Apparently they all stay. It is a common sight to see two or three aged men (for a Danish under-officer at 50 is aged) sitting on a baggage wagon smoking their pipes, and this must account for a good deal of the inefficiency observed. The under-officers have also a certain amount of political influence, and are generally able to get their own way.

FRANCE.

The following manœuvres were witnessed by British officers:—

(a.) Army manœuvres.
(b.) Manœuvres of the 2nd, 6th and 8th Cavalry Divisions.
(c.) Manœuvres of the 3rd and Provisional Cavalry Divisions.

(a.)

Army Manœuvres.

These manœuvres were divided into two periods as follows:—

First period—11th to 13th September (both inclusive).
Rest day—14th September.
Second period—15th to 17th September (both inclusive).

The distribution of troops during the first period was—

Blue army.—Commander—General Gallieni.
1st Cavalry Division.
10th Army Corps.
11th Army Corps.
Army heavy artillery.
Army air service.

Red army.—Commander—General Marion.
7th Cavalry Division.
9th Army Corps.
Provisional Army Corps (Provisional Division and 9th Division).
Army air service.

During the second period, the 9th Army Corps was transferred to the Blue army, and the Red army was reinforced by the 1st Reserve Division. (*See* page 46.)

Regular units contained 30 to 50 per cent. of reservists.

This year, billeting was arranged under service conditions and operations were continuous. Four or five years ago, manœuvres were often unreal, owing to an inclination to coddle troops; there was nothing of that sort in 1912.

Nature of the operations.—In the initial situation, both armies were on very broad fronts, with an average distance of some 55 miles between them. The Red force was on a front of about 75 miles, with its cavalry division on the northern flank. The Blue force had a front of some 48 miles, with its cavalry division in the centre.

11th September.—Both commanders decided to concentrate. General Marion, whose orders were to crush first the 11th and then the 10th Corps, made a forward concentration and narrowed his front to some 37 miles. General Gallieni concentrated on his

11th Corps by bringing the 10th Corps to the south by a flank march of some 20 miles. The opposing cavalry divisions did not come into serious contact.

12th September.—General Marion tried to delay the advance of the 10th corps by attacking it with a mixed detachment composed of the 7th Cavalry Division and an infantry (chasseur) brigade, while he made a forward concentration of the 9th, 17th and 18th Divisions against the 11th Corps. In spite of the admirable marching of the infantry of the mixed detachment, the attack on the 10th Corps was about 1½ hours too late; the march of the 10th Corps was not greatly retarded, and Gallieni practically completed his concentration as he intended.

To the south, the 1st Cavalry Division (Blue) delayed the advance of the 9th Division until 10 a.m. and then carried out a similar rôle against the 9th Corps. The 11th Corps got into touch with the 9th Corps and the former's advanced guards were driven back about 3.30 p.m.

13th September.—It was Marion's intention to hold the 10th Corps with the provisional division and to crush the 11th Corps with his three remaining divisions (9th, 17th and 18th). Gallieni intended to pivot on his left division and to attack in a N.E. direction with the remainder; the 21st Infantry and 1st Cavalry Divisions being used to turn Red's left flank.

The ensuing general engagement worked out in an interesting manner. At the beginning Red's right was slowly pushed back, but on the left the three Red divisions made good progress against the 11th Corps, the latter's right flank being turned by the 18th (Red) Division, while the 9th (Red) Division drove a wedge in between the left of the 11th and the right of the 10th Corps. Gallieni had, however, kept the 19th Division in reserve, and with this he delivered a heavy counter-attack against the right of the 9th Division. This was assisted by a sensational charge made by the 1st Cavalry Division against the rear of the 9th Corps, in which the corps artillery and four aeroplanes were captured, not to speak of General Marion and the corps commander, with their respective staffs. The action of the 7th Cavalry Division (Red) was indecisive.

This brought the first period to an end; the operations during the second period were not of such general interest.

(b.)

Manœuvres of the 2nd, 6th and 8th Cavalry Divisions.

These manœuvres took place from the 26th to 30th August in the departments of Haute-Marne and Côte d'Or. 72 squadrons, 18 guns, 18 machine guns and 3 cyclist companies were engaged.

Nature of the operations.—Operations, except on the last two days, took the form of daily schemes. Hours were fixed for the march of reconnoitring detachments and main bodies. The former were allowed at least 2 hours' start. The brigades usually marched at 5 a.m., and the squadrons detailed for exploration about 2.30 a.m., so that the latter were able to work methodically and without worry. The billeting areas—one for each division—were 12—15 miles apart.

The march of the various units to the divisional assembly point occupied from 1 to 2 hours, and was usually made at a walk. Reports from officers' patrols thus arrived soon after the division was assembled; that is to say, when the divisional general required them.

Protection was provided for the assembly of the division by squadrons in observation on a stationary line. Cyclists sometimes accompanied these squadrons, and for the first bound they generally preceded the division and seized a tactical point. In addition to these protective measures, divisions had their own advanced guards, usually consisting of one regiment, one battery and a brigade machine gun section.

Schemes were always based on a general situation with imaginary infantry columns, the duties of whose mounted troops were either to explore in advance, or to cover a flank movement or a retreat. Whatever the duty assigned to the cavalry, it was the rule that should the hostile cavalry be encountered it was to be attacked.

Operations consisted of an approach march and a mounted cavalry fight, and the procedure was always as follows:—

Découverte or distant reconnaissance by officer's patrols supported by contact troops or squadrons holding open passages on their line of retreat while troops detailed from other units were sent forward to furnish relays.

Couverture or protective reconnaissance by squadrons covering the assembly points and successive bounds of the main bodies.

Rapid occupation of advanced points of tactical importance by units with great fire power, *i.e.*, cyclists and machine guns.

Movements by bounds of main bodies covered by advanced guards. (This method of advance was not, however, invariable.)

Rapid issue by divisional generals of verbal orders to brigadiers and artillery commanders when the enemy's line of advance was ascertained.

Rapid and sudden development of shock action supported by fire from the artillery, cyclists and machine guns.

Deployment, delayed until the last moment, and the charge made with great solidity but without increase of pace (in peace time).

After the attack, troops stood fast and officers fell out to attend the conference. Units then marched to billets under peace conditions. For the mass, the exercises lasted 8—9 hours, viz., from the time of leaving billets in the morning to the time of return to billets in the afternoon.

(c.)

Manœuvres of the 3rd and Provisional Cavalry Divisions.

These manœuvres took place at the end of August in the area Beauvais—Clermont—Breteuil—Saint Just en Chaussée.

Each division was composed of three brigades of two regiments each, regiments being of four squadrons, and squadrons between 90 and 100 men strong. A *groupe* of two 4-gun batteries of horse artillery was attached to each division.

During the last two days an infantry regiment of three battalions took part in the operations.

Direction of Manœuvres.

Some criticism has been made to the effect that the Army Manœuvres afforded generals little opportunity of showing their strategical skill. Such criticism appears to be ill founded as the amount of latitude given to generals was precisely what would be allowed an army commander by his *généralissime*.* Manœuvres gave army commanders and their staffs an opportunity of directing large bodies of troops in grand tactics. Strategical schemes dealing with an isolated army of four or five divisions would have had no bearing on the problems of the Franco-German frontier.

Directing staff.-- The directing staff consisted of General Joffre, General de Castelnau, 4 aides-de-camp and some 25 other officers.

The aides-de-camp were all young staff college graduates, and their duties were numerous and varied. One of them was responsible for communications with the press, for the issue of the daily narrative and maps, and for the headquarter mess. One aide-de-camp was always on duty, day and night, and when in that position he was expected to answer any questions.

The most striking feature of the work of the directing staff was the absence of fuss, and the quiet and efficient way in which everything was done. No one seemed to get "rattled," no one seemed in a hurry, the machinery ran smoothly, and there did not appear to be a single hitch.

The directing staff was accompanied by an electric-light plant, capable of lighting three large rooms. This, dynamo included, was transported on one two-wheeled cart.

Umpires.—The umpire staff (including the directing staff) consisted of 62 officers, 40 cyclists and 113 mounted orderlies.

Every evening, as soon as the intentions of general officers commanding armies were known, an officer of the directing staff at once allotted zones to the various umpire groups for the following day, each group being intended, as far as possible, to follow the operations of both sides in a zone corresponding to that of an infantry division. This system had the advantage that umpires were thereby enabled to act as a combined organization without bias for any side or formation.

The following is an extract from the instructions to umpires :—

"It is within the powers of umpires to order troops to halt, but this power must be exercised in such a manner that the dash and offensive spirit of the infantry are not interfered with. As a

---

* The *généralissime* is the commander of the group of armies in any given theatre of operations.

rule, troops will be halted definitely by umpires only when officers persist in moving blindly forward under heavy fire without taking measures to make such an advance possible. Umpires should give their reasons for halting troops, and should not fix a definite time limit for the duration of the halt.

"Umpires will be guided in their decisions by definite facts such as the fire tactics of the unit, its formation, and the manner in which it crosses the zone of fire and makes use of the inequalities of the ground."

Damage to crops, &c.—All damages are assessed and paid for immediately by a commission, composed partly of officers and partly of civilians—the largest farmers in the neighbourhood.

Staff.

The headquarters of higher formations in the French army are classified as follows :—

1. General (*i.e.*, of the commander-in-chief).
2. Army (more than one army corps).
3. Army corps.
4. Divisional.
5. Lines of communication.

In the case of 1, 2 and 3, the headquarters consist of the general officer commanding, a chief of the staff, a *sous-chef* of the staff, and three bureaux. In the case of 1 or 2 (when the army is working independently), there is a fourth bureau, known as the *Direction des Chemins de Fer.*

The duties of the three bureaux are as follows :—

No. 1. Administration and supply.
No. 2. Information.
No. 3. Operations.

In the "Report on Foreign Manœuvres, 1911," page 33, it is stated that there is no separation of staff officers into groups, but that the practice is "to put all officers, except the *chef* and *sous-chef*, on a common roster to perform . . . whatever duty comes to them in turn." As regards divisions, latitude is given to the chief of the staff in allotting duties, but it is not the case in army corps or army staffs. The three bureaux given above issue separate orders, and officers in different bureaux do not interchange duties. (These remarks do not, of course, refer to special or occasional duties, such as *officier du liaison, officier de jour,* &c., which are undertaken by officers in rotation.)

Operation orders are issued in two parts. Part I., to which the 2nd and 3rd bureaux contribute paragraphs, is distinct from Part II. (administrative orders), for which the 1st bureau is responsible.

Army staffs.—No. 1 bureau (administration), in the headquarters of an army, is under the direction of a senior field officer,

---

who may belong to any branch of the service, but who must be a staff college graduate. The following are the principal matters with which this bureau deals:—

1. Consumption and renewal of supplies and stores.
2. Evacuation of sick, replacement of casualties, organization, strength, and provost-marshal's department.
3. All questions relating to departments (*services*).
4. States (rendered every 5 days).*

In addition to the foregoing three bureaux, there are two other groups of staff officers; but these groups, although belonging to the army staff, cannot be considered as part of the *état major* of the general officer commanding.

The first of these groups comprises the *intendant de l'armée* and his technical staff, principal medical officer and his staff, &c.; in short—heads of departments and their staffs.

The other group consists of the *directeur des étapes et des services* and his staff. This officer, who is officially known as the D.E.S., is a lieutenant-general, and assists the general officer commanding in all matters of administration. He exercises direct command over the *chefs des services* of the army; issues orders direct to them and deals with their requirements. He manages the entire system of requisitioning and exploiting of local resources. He works hand in glove with the *intendant de l'armée* as far as the technicalities of the latter's department are concerned. Together they submit proposals to the general officer commanding, who may or may not adopt them; but when the general has finally decided on his course of action, and has definitely settled his orders both as regards Part I. and Part II., then it is the duty of the D.E.S. to issue such additional orders (*ordres particuliers*) to his own subordinates as may be required. This is rendered indispensable by the fact that army orders only include information that it is strictly necessary for the army to know.

It is understood that the staff of the D.E.S. is divided into two bureaux. One bureau deals with—

Administration of area of operations.
Organization of the stages on the lines of communication of the army to which it belongs.
Establishment of magazines of supplies and stores.
Supplies for the troops and organization of *convois éventuels*.

It is the channel of communication towards the rear of the army.

The other bureau issues orders for movements and camping grounds, and is the channel of communication towards the front.

Army corps staff.—During the manœuvres there were in each army corps staff, 6 officers in No. 1 and a total of 8 officers in the

---

* On every 5th day, commencing from the 1st of each month, commanders of units render through the usual channels states which, finally tabulated by army corps, find their way to the staff of the supreme commander, who forwards them to the War Minister.

other two bureaux. The duties of the junior officers varied each day. These duties were—

Officier du jour.
  „ *de piquet.*
  „ *de ravitaillement.*
  „ *de liaison* (one to each division or corps in the same army).

The *Officier du jour* opens letters addressed to the general officer commanding; keeps a record of all incoming and outgoing information whether verbal or written, correspondence and orders given or received; regulates and supervises the secretarial staff; remains by the side of the chief of the staff in the field, and by night sleeps in the headquarter office.

The *officier de piquet* replaces the *officier du jour* if required. He is responsible for the personnel and *matériel* of headquarters on the march.

The *officier de ravitaillement* is present at the refilling of regimental trains; is responsible that times are adhered to and that the traffic is regulated; settles disagreements between issuing and receiving parties.

The *officiers de liaison* maintain communication with the units to which they are detailed for the day (through them orders are transmitted), and afford commanders such additional information as may be useful to them concerning the intentions, &c., of the general officer commanding. On return to their own headquarters they convey information regarding the unit they have just quitted to the army corps commander.

At army corps headquarters there are, in addition to the above, the Intendant-General and a staff of some 15 officers. Although there is no *directeur des étapes et des services* on the staff of an army corps, it cannot be said that the Intendant-General takes his place.

In the case of the cavalry divisions, the staff was expanded for the manœuvres by the inclusion of reserve officers for the work of information and administration. These same officers perform the same duties every year.

The staff of a cavalry brigade consisted of the brigade major and a captain from the reserve, who performed work similar to that of our brigade staff captains.

Uniform.—The system of staff arm-bands (blue for brigades, red for divisions, tricolour for army corps, with the number of the formation in each case) makes it easy to identify the various headquarters.

Quality of staff work.—The ease and certainty with which masses of men are moved in any required direction is good evidence of the thorough efficiency of the staff officers. The following are examples of what was done :—

On the 11th September, two divisions marched 20 miles on one road, the line of march being crossed early in the day by a cavalry regiment.

On the 15th, the 7th Cavalry Division was brought across from the right to the left flank of the Red army, crossing the line of march of the three columns of the provisional division, which was engaged at the time.

On the 16th, the provisional division was relieved on rear guard and sent through the 9th Division to form a general reserve in rear.

Two points were noticed—

First, the *trains regimentaires* were kept some miles behind the fighting line. This naturally kept the road very clear, but on the other hand the men did not see their wagons until 7 or 8 p.m.

Secondly, this class of staff work is of comparatively recent growth. A few years ago, when one division was marching on one road it was necessary to detail a staff officer to regulate the march: now, two divisions are marched on a road without any such special measures being taken.

A characteristic of French staff officers is the way they train the memory. They are not dependent on note-books. At conferences, generals usually state the case for their side without referring to their notes; in the same way they are not always looking at their maps but get the country fixed in their minds. There is a legend that General Tremeau, could direct a big staff ride on the German frontier without a map.

Orders.

Orders were noticeably shorter and more concise than was the case five or six years ago.

The following orders were issued by army commanders:—

1. " *Instruction Générale, Personelle et Secrète*," issued to general officers commanding army corps and cavalry divisions only. These were issued from time to time as required, not necessarily every day,

2. " *Ordre Génerál d' Opérations*," in two parts; the first dealing with operations, the second with administration. The two parts may be issued at different times. Officers receiving Part I. do not necessarily receive Part II., but nothing definite is laid down on this point, which is left to the discretion of the chief of staff. In practice, more copies usually are issued of Part II. than of Part I.

3. " *Instruction spéciale à la division de cavalerie*." Usually only issued during the approach march.

4. " *Orde de stationnement*."

5. " *Instruction particulière pour les escadrilles*."

6. " *Instruction particulière au dirigeable*."

In addition to these an intelligence bulletin usually was issued every evening. It is not known to whom the latter was sent.

Corps commanders issued Nos. 2 and 4 of the above, and if necessary No. 5. In addition, both corps and subordinate commanders issued an *Ordre préparatoire* giving the hour of starting and order of march. The general officer commanding cavalry division occasionally issued *instructions à la découverte.*

Spare copies of operation orders usually were issued. The number is not laid down, but the principle is to keep the higher commanders *au courant* of the situation in case they may have to assume command of a unit higher than their own. Thus the corps commander receives a sufficient number of army orders to give a copy to each divisional commander, the divisional commander enough for each of his brigade commanders, and so on.

The system of maintaining communication with the headquarters of a force on the move is as follows :—

In orders daily, two places are named to which communications for headquarters should be addressed. The first (A) generally being the place where headquarters are spending the night, the second (B) being the place where they expect to spend the next night. In orders it is laid down that headquarters will be at A until a certain hour, and from that hour they will be at B.

A staff officer, with motor cyclists, remains at A until the appointed hour. A second staff officer is sent on by motor with other motor cyclists, telegraph section, &c., so as to be installed at B at the hour named, when the staff officer at A packs up and rejoins his headquarters.

Orders for resuming march in one column from billets.—The corps commander states in his operation orders the hour at which the heads of the following formations should pass the starting point :--

1. Main guard.
2. Main body.
3. Corps artillery.
4. Rear division.
5. Regimental train (*trains regimentaires*).

Based on these, the commandants of the various billeting areas (who, in an army corps, are usually the two divisional commanders and the senior officer of the corps troops), fix their own starting points and hours of march.

"24 *o'clock.*"—This year time was reckoned in orders from 0 to 24 o'clock, viz., from midnight to midnight. The system is a success; it is quicker to write, and eliminates at least one possible source of error.

Tactics of the three Arms combined.

Two points were especially noticeable in the tactical handling of troops, namely, the enterprising and aggressive spirit animating all ranks and the power that the French have acquired of fighting *encadré.*

Allusion is sometimes made in England to a "French system" of tactics, the bases of which are the employment of general advanced guards and general reserves. The term "French system" is a misnomer, as although certain French military writers have advocated the above doctrine, it has never received the authority of the French General Staff. The latter, although they

have not opposed the use of general advanced guards and general
reserves under certain conditions, have certainly never laid down
that the secret of successful war is to be found in this or any other
"system." During the 1912 manœuvres a general advanced guard
was not employed by either commander.

General reserve.—A general reserve was employed on four
occasions, the method of employment being different in each
case.

1. General Marion's orders for 14th September laid down that
the 18th Division was to be in a certain zone and was to constitute
the army reserve, its mission being to turn Blue's right flank;
that is to say, the position and rôle for the general reserve was
fixed before the action definitely began. There was nothing in
orders to show whether the responsibility for committing the
18th Division was to rest with the army commander, or with
the general officer commanding the 9th Corps to which it belonged.

2. On the same day the 19th Division practically formed a
general reserve to the Blue army, although this was not stated in
army orders. It was kept in reserve some $2\frac{1}{2}$ hours longer than
the Red general reserve and came into action when the right
flank of the 9th Red Division became exposed during its efforts
to turn the left flank of the 22nd (Blue) Division.

This was an interesting example of a general leaving an
intentional gap between two portions of his army in order to
induce his opponent to give him a flank on which to fasten.

3. During the second period, General Gallieni (Blue) kept the
22nd Division in reserve until the evening of the 16th September.
He then took it from the general officer commanding 11th Corps
and handed it over to the 9th Corps, to be used in conjunction
with the 17th and 18th Divisions for the decisive attack.

4. During the second period, General Marion covered his
retreat with the provisional division. Having crossed the line of
the Creuse and the Vienne, he then relieved this rear guard with
fresh troops on the 16th and formed it into a general reserve at
L'Etang. He intended to use this force for a decisive counter-
stroke on the 17th, but it is believed that the attacks of the Blue
force were so incessant that the general reserve had to be used
for stopping gaps and not for striking.

Local reserves.—These were small. On 13th September, the
commander of the 22nd Division, who was fighting a "backwards
and forwards" action all the morning against the 17th and 9th
Divisions, only kept one battalion (out of 12) under his own
control.

If opportunity occurred, troops were brought back from the
firing line to form a local reserve. On the occasion above mentioned, the 22nd Division was slowly pushed back by the superior
forces opposed to it until it finally came to a stop on a high commanding ridge. While the enemy were preparing for an organized
attack on this ridge, the commander of the 22nd Division withdrew
a battalion from the firing line to strengthen his local reserve.

Protection.—The protective detachments which at one time
used to precede the larger formations on the march were con-

spicuous by their absence. Apart from an advanced guard and, occasionally, a flank guard, columns were kept together.

When halted for the night, in addition to ordinary outposts, columns sometimes protected their flanks with detachments (usually of a battalion and a battery). Thus on the night of 11th September, the 11th Corps had two of these detachments holding the bridges at Availles and Gourge, $1\frac{1}{2}$ miles and 4 miles respectively from the billeting zone of the corps. The present tendency, however, is to reduce these detachments to a minimum.

The normal strength of other protective detachments was as follows :—

Advanced guards :—

Army corps	1 divisional squadron. 1 infantry regiment (3 battalions). 1 artillery brigade.
Cavalry division ..	1 regiment of cavalry. 1 section horse artillery. Brigade machine guns. 1 cyclist company.
Rear guard to army corps in an advance	1 infantry company.
Flank guard to army corps during flank march ..	1 infantry brigade (6 battalions). 1 regiment of cavalry. 1 brigade of field artillery. 1 engineer company.

Night operations.—Troops in touch with the enemy were not allowed to engage in night operations without the permission of the Director. This was given only on one occasion.

Columns were allowed to march at night, and frequently did so.

Fronts.—On the 13th September, the three Red divisions engaged in the decisive attack were on a front of $6\frac{1}{2}$ miles, or rather less than $2\frac{1}{4}$ miles per division. In the same action, Blue had four divisions on a front of about 10 miles, and the provisional Red division, which was making a holding attack, was on a front of $4\frac{1}{4}$ miles.

On the 17th, the 22nd Division near Civray, made a decisive attack on a front of little over 1,000 yards; this was, however, the "Presidential set piece," and was hardly a normal example of French tactics.

Marches.—On nearly all occasions each corps (two divisions) marched on two roads, the distribution, itinerary, and hour of starting being fixed by the corps commander.

At the beginning of the second period, when the Blue force had a pursuing rôle, the centre corps (11th) marched on two roads (the leading brigades in each column both belonging to the 21st Division); the 9th and 10th Corps each marched on three or four roads in order to press the hostile rear guard as quickly and closely as possible.

Infantry marched in fours, closed up, and to the side of the road. At a halt, of even short duration, arms were piled on the

edge of the road; the valise was taken off (with one hand) and
placed under the pile, and the men sat down clear of the road.
Cavalry, in half-sections, moved on both sides of the metalled
portion; cavalry divisions, however, when on the road, marched
in sections.   Artillery always halted well to the side of the road
with riding horses backed in between guns, heads towards the
road.

The order of march of a column, consisting of the 41st Brigade
and headquarters of the 21st Division, is given below.   The
army was then advancing in several such columns following up
the retiring Red army, with whom contact had not yet been
re-established.

> Divisional mounted troops, 1 squadron Chasseurs.
>   Interval of about 2 kilometres.
> Two companies, 137th Infantry Regiment. ⎫
>   Interval of about 1 kilometre.   ⎪
> 1½ battalion, 137th.   ⎬ Advanced guard.
> Company of Engineers.   ⎪
> G.O.C. Division and Headquarters.   ⎭
>   Interval of 1 kilometre.
> Remainder of 137th.
> One group of 51st Artillery.
> 93rd Infantry Regiment.
> Remainder of 51st Artillery.
> Baggage of Chasseurs.
>   „   „ Artillery
>   „   „ 137th Infantry.
>   „   „ 93rd Infantry.

The baggage formed a short column, and there was no interval
between it and the artillery.

Entrenched position.—During the second period, the Red 54th
(Reserve) Division entrenched a position covering the passage of
the River Vienne.   The main line of resistance was 10½ kilometres
in length, parallel to, and from 1,500—200 yards from the river,
and protected by a bridge head (garrison—1 battalion) at Pouzay,
and by a detachment (1 battalion) between Maillé and La Celle.
A second line, about 4 kilometres in length, was entrenched
1,400—1,900 yards in rear of the main line.   A front of about 2
miles was allotted to each brigade.   Brigades were distributed
thus—

> One regiment—firing line and supports.
> One regiment—local reserve.

The advanced detachments were supplied by the regiments
finding the firing line.

The Division furnished the working parties, who were assisted
by 2 companies of engineers, the Corps Engineer Park and
1 section of the Army Engineer Park.

A reliable authority stated that all the trenches were sited
by the engineers.   In any case they were well sited, well masked
and concealed, with no sky line behind them and a good field of
fire for 600—900 yards; they followed the contour of the ground

well, and flanking fire was provided for. Each trench accommodated about a section : a company (about 150 rifles) held a front of about 800 yards.

Concealment was evidently the chief factor, both in the siting and construction of the trenches, and it certainly would have been difficult for the Blue artillery to locate them. In order to make the trenches as inconspicuous as possible, no head cover, overhead cover or traverses, were made.

The position was apparently designed with the intention of inducing Blue to attack the eastern flank, thereby exposing himself to a decisive counter-attack from the neighbourhood of Draché.

The officer commanding the bridge head at Pouzay stated that the battalion forming its garrison would, in case of attack, be reinforced by a battalion from the nearest trenches, to which both battalions would retire when forced to evacuate the bridge head.

Infantry.

Battalions were between 700 and 800 strong, from half to one-third being reservists.

Regiments had been marching from 6 to 10 days before manœuvres began ; these marches had caused few casualties (one regimental commander stated that he had only four men sick).

The marching was certainly very fine. On the 12th September, the Chasseur Brigade started at midnight, and when seen at 9.30 a.m., were moving at very nearly 4 miles an hour. They marched uninterruptedly (except for the hourly halt), until 11 a.m., when they came into action and continued fighting till dark. Their commanding officer put the distance marched at 40 miles (5 or 6 miles of it across fields), and this was not an over-estimate. On the 15th September, the Colonial Brigade started at midnight, and when seen at about 4 p.m. were slowly doubling along a road after some cavalry. After dark, a particular request was sent to the Director, asking for leave for them to continue the action after nightfall. It is only fair to say that the Colonials had no reservists and the Chasseurs very few ; units with reservists, however, marched extremely well, a march of 30 to 35 miles being nothing extraordinary.

In spite of the recent improvement in marching in the British Army, the standard is still distinctly below that in France. This may be due to the following reasons :—

1. In certain corps the infantry march a minimum of 5 miles fully loaded *every day*. One division has to march 9 miles to its training ground and 9 back, in addition to the daily work.

2. Billeting enables men really to rest. With the British system of bivouacking troops in September, it frequently happens that the men are too cold to sleep. This saps their vitality, and it is impossible to expect them to march as well as men who have enjoyed several hours genuine rest.

Tactics.—Once they were deployed, the French infantry displayed marked inferiority to our own in minor tactics. There was not the same dash nor anything like the same efficiency in fire direction and control. Like the cavalry, the infantry did not seem to realize what modern rifle fire is like.

It is true that at battalion training this is not so much the case; then, practical formations are adopted for supports and reserves; men lie down instead of kneeling, and more attention is paid to cover. The truth may be that at army manœuvres the infantry do not take the trouble to " go through their tricks."

The attachment of cavalry scouts to infantry still continues, 18 being allotted to each infantry brigade (6 per regiment and 6 for the brigade commander). The men are reservists undergoing training, so that the inconvenience from a cavalry point of view is thereby lessened. The infantry commanders greatly approve of this system, and appear to make considerable use of the men for scouting purposes.

One company was seen entrenching in the attack, one man digging with his entrenching tool from a lying-down position while his neighbour covered him with his fire. The entrenchment made was very small—about a quarter of an hour's work.

Volley firing was in many cases employed, and companies were seen firing volleys with closed ranks. As has always been the case, bayonets were fixed, swords drawn, cavalry charged through infantry, and colours were carried in the assault.

Machine guns (infantry).

Although cavalry machine guns were frequently brigaded, no instance of this was seen in the infantry.

No new developments as regards tactics were seen. The most successful methods of using machine guns with infantry appears to have been for the infantry commander to tell the machine gun commander his intentions, and perhaps on which flank he wished the machine gun to operate, and leave the execution of this task to the machine gun commander.

In the "line of battle fighting" that must occur when large masses are in action, it is difficult for machine guns to work on the flanks of their unit. The French solved this difficulty by sending on a proportion of their guns with the firing line and placing others on high ground to fire over the heads of the infantry. This was well carried out.

Infantry machine guns were always worked in pairs.

Experimental sections of machine guns on cycles were attached to certain regiments for the manœuvres. The section consisted of a couple of the usual machine guns, each gun and mounting being divided into three parts, and so carried on three cycles under the top bar of the frame. The section carried 10,000 rounds packed in boxes each containing 300 rounds, two such boxes being carried on a cycle. The rounds were put up in metal strips, each of 25 cartridges, and the strips fed into the gun one after another. With one cycle carrying accessories and another for the lieutenant, the section comprised 24 cycles in all. It certainly can get forward under cover where pack horses cannot go, and though the weight (50 kilos.—110 lbs.) makes the cycles heavy to ride, they can move more rapidly than pack horses. The experiment was successful as far as it went, but it was not a severe test—the weather was fine, and there was little cross-country riding.

Cavalry.

Organization.—Cavalry divisions consisted of three brigades of two regiments each (total 24 squadrons) and two batteries of horse artillery. A cyclist company also was allotted to each division, except the 3rd and Provisional Divisions.

Tactics.—The initial dispositions of the cavalry on the Army Manœuvres were interesting.

The Blue cavalry were given a double rôle, namely, to discover the strength and direction of march of the hostile forces in three given areas, and to protect the left flank of the 10th Corps. The commander of the 1st Cavalry Division interpreted these instructions as follows :—

The division made a first "bound" of about 20 miles. In addition to the ordinary protective detachments, the cyclist company was detailed to follow the advanced guard for some 8 miles and then to move rapidly forward, seize the bridges over the River Dive, and form a screen protecting the division at the end of the initial bound.

Distant reconnaissance was carried out by means of one contact half-squadron supporting five officers' patrols, and a sixth independent patrol. The officers in charge of patrols were detailed by name in divisional orders; short instructions to each of them were embodied in the *Instruction à la decouverte*, issued by the divisional commander, which contained orders for 2 days, and the positions to which orders were to be sent for 3 days.

The rôle of the Red cavalry was simpler as they had only to discover the movements and strength of the 10th and 11th Corps. In this case four officers' patrols were sent out, supported by two contact squadrons. The patrols were to stay out 3 days if necessary.

The corps cavalry were reduced from a brigade to a regiment, but were nevertheless used for distant reconnaissance rather than protection. In the Red army, General Marion gave the corps cavalry a mission analogous to that of the cavalry division, each being allotted a sector for distant reconnaissance. The corps cavalry were ordered to be ready to assist the cavalry division if required, but they were not put under the divisional commander.

A distinct feature of French tactics is the employment of mobile infantry, with cavalry. Both this mobile infantry and the cyclists are intended to give an increased fire power to the cavalry and to relieve them of dismounted work.

At one of the conferences the Director laid stress on the fact that this infantry was not intended as a support for the cavalry, and must not be used as such. Cavalry was supported from in front, not from the rear; their support must be the enemy, and he thought nothing was so destructive to the moral and scope of a cavalry division as an infantry support. The infantry must work.

with the cavalry, and each must be made independent of the other. To this end a proportion of cavalry (and when necessary, artillery) was allotted to the infantry.

Theoretically the cavalry do not admit that they require infantry support; practically the fact that an infantry support is available gives them confidence.

A point of interest as regards the co-operation of infantry and cavalry occurred on the 15th September, when an infantry battalion was sent by mechanical transport to the assistance of the 1st (Blue) Cavalry Division. The vehicles were those that had delivered supplies to the *trains regimentaires* and were consequently empty.

The French cavalry continue to exhibit great contempt for rifle fire. Dismounted action is rare.

Horsemastership.—The horsemastership was surprising. The horses were quite fresh at the end of the manœuvres, although they were on their legs daily from dawn until dark. This is remarkable at the end of a hard training season, especially in view of the facts that the men never dismount, that the cuirassiers carry 18 lb. extra dead weight in armour plating, and that no feed is carried on the saddle.

One reason for this desirable state of affairs is that the system of billeting at least gives a horse shelter from wind and often a roof over his head.

March discipline.—The march discipline seemed to be better than our own. Columns kept well to the side of the road, and there were no lost distances. The head of a column always moved at an even pace and never more than 8 miles an hour; there was therefore no necessity for the troops in the rear to gallop to close up with those in front. Checks did not occur frequently, but when they did, they were signalled back along the column by all those who saw what was happening. There was practically no falling out, no "roughing" of horses, and no slovenly riding.

Officers' patrols.—The officers' patrols usually consisted of one officer and five men, and they were expected to remain out for two or three days if necessary. They covered considerable distances—one was met about 11 a.m. which had come 45 miles that morning—and were led with good judgment and great enterprise. Men were sent back singly with messages, and in spite of the fact that they had no map, they were said to get back quickly and safely.

Officers.—In spite of their age (squadron leaders under 40 are rare), the activity of the officers is remarkable, and a stout officer is unknown. They are thoroughly up to their work, very quick at making a decision, and full of dash.

The French cavalry is animated by a thoroughly patriotic spirit. All officers understand and subscribe to their doctrine and theory, which are clear and simple, both in tactics and training. Any half-dozen French cavalry officers would solve a tactical problem on the same principles. This uniformity of thought is probably due to the system of corps promotion and to the training at the cavalry school at Saumur.

French cavalry officers are better mounted than our own. Their chargers show more quality, have better paces, and are better trained.

Men.—The men are of good physique and ride well. They take an interest in their horses. Their discipline appears good, and they are cheery and willing. There is a large number of re-engaged men in the ranks.

Arms.—All heavy cavalry now carry the lance and the arming of hussars and chasseurs with lances has already begun. It has been decided that the cavalry soldier shall carry a bayonet and one will be issued as soon as the type has been selected.

The carbine is carried by the cuirassiers in a bucket on the near side of the horse; by the remainder of the cavalry it is carried slung on the back. In the latter case the sling passes over the left shoulder and the butt of the carbine is held steady by a clip attached to a leather waistbelt.

No bridging material was brought to manœuvres. It was stated, however, that each cavalry regiment carries in one wagon two metal boats. There is no brigade or divisional bridging material.

Equipment, saddlery, &c., on man and horse.—The following are carried on the man :—Carbine (slung), ammunition in pouches, water-bottle.

On the horse are—

> One pair of wallets (containing one suit of slacks, a pair of slippers and any remaining rations).
>
> Overcoat (rolled long and thin behind the saddle, and steadied to the saddle girth by a tab).
>
> Head-rope.
>
> Corn-sack (over the wallets).
>
> Forage cord (on off wallet).
>
> Blanket (under saddle).
>
> Sword (on near side).
>
> Canvas bucket (one per two men).

No picketing pegs are carried, a double shackle above the hocks being used for kickers. One *marmite* (or camp kettle) per troop is carried on the squadron *fourgon.* Each regiment has its own butcher's implements.

Each squadron has a colour which is painted on lances, saddles, wagons, &c. In addition, the horse's name is painted on the back arch of the saddle. Light double bridles of various patterns are used. Some of the snaffles are made in two pieces, like our own bridoons, some in three pieces. Some of the bits have half-moon ports, some have low ports, or other variations. The bits are lighter than ours and, since squadron and troop leaders have a variety to choose from, they fit better. Consequently horses carry their heads much steadier and their paces are smoother than is the case in British regiments.

Transport.—The transport of a cavalry brigade comprises—

	4-wheeled, 2-horsed.
Fourgons for baggage, *i.e.*, officers' kits, field forge, *marmites* (4), squadron books, &c. ..	10
6 wagons per regiment for *alimentation* (forage and rations)	12
Brigade headquarter wagon	1

	2-wheeled.
Telegraph cart	1
Ambulance carts	2
Total	26 vehicles.

With two machine guns, the total number of vehicles in a brigade is 28. There are no travelling kitchens in the cavalry.

Inter-communication.—No wireless was used, and there are no visual signallers in the French cavalry, but each brigade has a cable detachment of one non-commissioned officer and eight men with 16 miles of wire in a two-wheeled cart. The duty of this detachment is to keep brigades in touch with the divisional commander and with other brigade headquarters.

In the field, messages are transmitted by officers and orderlies, or use is made of the existing telegraph and telephone lines.

The orderlies are well trained, and are generally long service re-engaged men. The French consider that cyclists and motor cyclists are so easily captured that they are too unreliable for inter-communication.

Machine guns (cavalry).

There are two machine guns in each brigade. The detachment, which usually is attached to one of the regiments for rations, billets, &c., consists of one officer, one non-commissioned officer, and 30 men, all of whom are mounted and carry revolvers but no rifles. The guns are made on a system which has been evolved from the Puteaux and Maxim systems. They are clip loaders, each clip holding 25 cartridges. Blank cartridges have hollow wooden bullets which allow of the necessary recoil; consequently, the action for firing blank is the same as for firing ball.

Each gun is mounted on a steel carriage which is two-wheeled and four-horsed. This carriage, which is light and mobile, carries 14 ammunition boxes containing 3,500 rounds, two spare barrels for the guns, one pick, one felling axe, and one shovel. In the divisional ammunition column, there are 18,500 rounds for the machine guns of each brigade.

The gun can be fired either from the carriage or from the tripod on the ground, but the latter is the normal method. The gun is permanently fixed to the tripod, and can be brought into action on the ground in 30 seconds; it has an all-round traverse, and has no shield.

issue supplementary orders is when the corps artillery and one or both regiments of divisional artillery were to be massed under one command.

In a division the senior artillery officer on the divisional staff is a colonel, and might often be junior to the colonel commanding the divisional artillery regiment. This would make a further complication as regards command. As a matter of fact, the duties of the senior artillery officer on both army corps and divisional staffs are administrative (ammunition, supply, &c.), rather than executive.

The command of mixed forces.—When a force of artillery and infantry were given a definite task, such as the attack of a ridge or village, General Percin was an ardent advocate of putting the artillery under the direct command of the infantry officer. There was always considerable opposition to this view on the part of artillery officers, and since General Percin's retirement the system has been discontinued, because in practice it was found impossible to get the artillery back. The infantry commander having gained his first objective, then attacked the next, and not unnaturally ordered his artillery to support him. The consequence was that the artillery slipped away completely from the control of the divisional general.

Artillery with cavalry.—Stress is laid on the fact that in cavalry actions the artillery must come into action on the forward slopes and use direct fire, with the enemy's cavalry, and not his guns, as its objective. This appeared to be done in every case, and it was only when the provisional division was endeavouring to force the passage of the Brèche, on the 29th September, with its two battalions of infantry that the artillery came into action in covered positions. In every cavalry encounter the guns of one side or the other got more or less masked by their own cavalry. This was due to a variety of causes—insufficient information, miscalculation of time and space, or inability of the guns to keep up on the heavy ground. On the second day, the artillery commander of the provisional division had to confess that his fire was completely masked by his own cavalry after the first salvo, whereas the opposing guns were able to fire effectively up to the actual moment of contact at a range of 1,500 yards. As the Director said, this may have been due to bad luck. The division was operating well away from its guns, but unfortunately was obliged to wheel outwards to meet the enemy, who were found in a rather unexpected direction. As long as the cavalry can circle round the artillery, he said, all is well; but as soon as the cavalry turns away, the danger of masking the artillery at once comes into view, and the latter must shift its position with all possible speed.

A favourite manœuvre appeared to be for one of the two divisional batteries to be thrown well forward in observation, while the other remained limbered up with the cavalry until the tactical situation became clear, when it galloped forward. If isolated, the artillery was given an escort of a squadron.

On two occasions it was noticed that only three guns of a battery were firing; the fourth remained in observation, watching some approach where a movement was expected.

Observation ladder.—One observation ladder and wagon accompanied each field artillery brigade (*groupe*).

Heavy artillery.—The following units were present at the army manœuvres:—

1 brigade 155 mm. (6·1-inch) howitzers (experimental).
1 battery 120 mm. (4·7-inch) guns.
1 battery 106·7 mm. (4·2-inch) Creusot (experimental).
1 brigade 220 mm. (8·6-inch) mortars.

The 120 mm. gun was of a very old pattern (date 1878), mounted on an overbank carriage. The platform was heavy, and the work of making the anchorage for the hydraulic buffer was considerable. It took about 4 hours to lay the platform. An observation ladder was used, and a gyn was provided for shifting the guns from the travelling to the firing positions.

The 220 mm. mortars were also of an old pattern (date 1884). The mortar, carriage and platform were each transported on an 8-horse wagon. It took 1¼ hours to lay the platform and mount the mortar, and about the same time to dismount it. A gyn was provided for mounting the mortars and their carriages. Observation was by ladder. The bursting charge of the shell was 40 kilogrammes (88 lb.) of melinite.

Air Service.

The following aeroplanes were employed on the army manœuvres:—

Blue Army.

Escadrille I ..	..	6 two-seater Henri Farman.
,, II ..	..	6 two-seater Henri Farman.
,, III ..	..	6 two-seater Blériot,
,, A ..	..	6 monoplanes (3 Borel, 3 Blériot).

Red Army.

Escadrille IV ..	..	6 two-seater Deperdussin.
,, V ..	..	6 two-seater Maurice Farman.
,, B ..	..	4 Hanriot monoplanes,
,, Mixte	{	2 three-seater Deperdussin.
		2 three-seater Breguet.
		2 three-seater Nieuport.

In addition to these 46 aeroplanes, another *escadrille* of 6 machines was employed with the artillery, and 8 more arrived at various periods during the manœuvres, bringing the total number up to 60.

Organization.—The aeroplane service is organized as follows:—
(*a*) *Campement des escadrilles*, (*b*) *parc d'aviation*, (*c*) *reserve de ravitaillement* (on the railway).

The distances between these echelons were naturally variable. At one period the *campement* was about 13 miles behind the infantry outpost line, the *parc* was some 2 miles further to the rear and the *reserve* was some 5 miles behind the *parc*. None of these units had moved for 2 days.

Reconnaissances were made either direct from the *campement* under the orders of the army commander or an *escadrille* was attached to one of the corps or to the cavalry division. In the latter case, the officer commanding preceded his *escadrille* by motor in order to select a suitable landing place. At the close of the day's operations, the *escadrilles* always returned to the *campement*. Landing places were selected as close as possible to the headquarters of the general officer to whom reports were to be given.

The *escadrille* attached to a cavalry division was almost always a monoplane unit.

Nature of employment during manœuvres.—So far as is known, aeroplanes were used exclusively for reconnaissance; they possessed no material for offensive action.

On occasion, particularly when forces were distant, *escadrilles* were lent to various commanders. Thus, on 11th September, the Red Commander-in-Chief lent one *escadrille* each to the commanders of the 7th Cavalry Division, 9th Corps and Provisional Corps. On the 12th, the command of these units was re-assumed by the Red C.-in-C.

The system of scouting at the beginning of manœuvres was to divide the area up into zones, each zone being allotted to one *escadrille*. Theoretically, a patrol went out every 2 hours (though this was not always the case); if a hostile body was observed in a zone, its strength and the direction of march were noted. It was then the special duty of the next patrol to locate this force with certainty, in addition to its ordinary work. This system provides a basis for continuity in reconnaissance, and may defeat concealment tactics on the part of troops.

As the two armies approached each other and the situation became more defined, the zone system was modified according to circumstances.

On the Blue side, observations by aeroplane were usually made from a height of from 500–800 metres (1,640–2,625 feet); by airship from a height of 1,300 metres (4,265 feet).

When clouds were low, aeroplanes kept at the edge of the clouds.

Range of reconnaissance.—The Red aeroplanes averaged about 100 miles per journey, the Blue about 165. It is said that General Marion restricted the range of his aeroplanes because he was anxious to get his information back early. Whether this was the case or not, there is no doubt that the information obtained by Blue was distinctly superior to that of Red.

Results of reconnaissances.—The following notes on the official intelligence reports show the nature of the information obtained :—

(*Blue.* 11th September. 3 p.m.) The points reached by the Chasseur and Colonial brigades at 9.30 a.m. were correctly reported; they were said to have been accompanied by two brigades of artillery (*one* was correct). No other information was published.

(*Red.* 11th September. 5.30 p.m.) The 11th Corps was reported to have billeted some 5 miles west of its real billeting zone.

A division was reported between Thouars and Montreuil; this was in reality an army corps less one infantry and one artillery brigade.   These two units, which were acting as flankguard, were correctly located, but were described as *petites fractions d'infanterie.*

It is interesting to notice that one aeroplane reconnaissance reported that no troops had been seen on the Thouars—Montreuil road at 12 noon.   At this time very nearly an army corps was marching along this road, the head of which left Montreuil for Thouars at 8 a.m.

The cavalry division was located at 1 p.m., but its direction of march was reported in a misleading way.   There was no mention of it after 1 p.m.

On 13th September, the 1st " Blue " Cavalry Division was located by the Red airmen at Thénezay.   It was then lost until it reappeared some 3 hours later 10 miles from Thénezay in rear of the Red army, where it made a successful and sensational charge.

On 15th September, the " Dupuy de Lôme " (Blue), reported that troops (number unknown) had detrained at St. Maure, and that a defensive position had been organized south-west of that point (no details or limits given).

On the evening of the 15th, Red located the centres of the billeting areas of the Blue columns as follows:—10th and 11th Corps respectively 5 and 3 miles north-east of their real positions, the 9th Corps 5 miles south-east of its true position.

The fact that the Blue cavalry had been supported by an infantry battalion conveyed by motors was reported.

The conditions were extremely favourable.   The weather was good; there was little wind (except on two mornings); the atmosphere was cool; the country was very suitable, being practically one vast aerodrome, on which landing and starting was everywhere possible ; and as nothing was to be feared from hostile aircraft or troops, both sides may be said to have had command of the air.   It would be a fortunate army that enjoyed all these advantages in war.

A proper appreciation of the work carried out by the aircraft cannot, naturally, be given without access to all reports, and without knowledge of the times at which information was received.   As shown in the intelligence reports, there were a good many gaps in the information.   Taking the results generally, the information obtained may be said to have been valuable, but by no means exhaustive.   The following points were brought out:—

1. The necessity for trained observers.   Reconnaissance by untrained officers was said to be almost valueless.   Of the trained observers, staff officers usually got the best results.

2. The difficulty of accurately locating a large unit when it has shaken out into billets.   The billeting area of a division was seldom, if ever, correctly given.   Camps or bivouacs would obviously be much easier to locate and estimate.

3. Distant reconnaissance during the approach march is likely to give better results than close tactical reconnaissance. In the former case the direction of roads, clouds of dust, &c., to some extent guide the observer, while it is easy to estimate the size of a column on the march. Once troops leave the road they can be more easily hidden, and the pilot finds it more difficult to effect a compromise between flying high to avoid fire and flying low to observe better.

4. Observers will find it difficult to avoid "losing" rapidly moving troops, such as cavalry or cyclists.

Organization of an escadrille.—*Escadrilles* had the following establishment :—

Matériel.

6 aeroplanes.
6 motor wagons (*fourgons* or *tracteurs*).

Personnel.

Commanding officer	1
Pilots	6
Observers	6
Mechanical transport drivers	6
Mechanics	6
Various (cooks, &c.)	6
Total	31

In war, two reserve aeroplanes would be included, which would necessitate an increase of 10 to the personnel and of 2 wagons to the transport.

The motor wagons (weight about 2,500 kilos. or 5,512 lbs.) have pneumatic tyres, those on the rear wheels being twin. They carry oil, motor spirit, minor spare parts, spare propellers, aeroplane tents, first-aid boxes, &c. At the rear of the wagon is a limber-hook to which a two-wheeled pneumatic tyred trailer can be attached. The trailer weighs 350 kilos. (722 lbs.) unloaded, 800 kilos. (1,764 lbs.) loaded, is about 16 feet long, and can travel at 25 km. (40 miles) per hour; its wheels are much nearer the rear than the front end. Both Blériot and Deperdussin monoplanes were seen packed on these trailers and occupied very little space.

Accommodation.—The Maurice Farmans are left out in the open at all times, their wheels being sunk to the axles in small trenches and the planes secured by ropes and pickets. The Henri Farmans are left out in fine weather only.

The monoplanes are housed in green canvas tents, supported by tubular steel posts about 9 feet high, the whole weighing 250 kilos. (551 lbs.) The tents occupy little space when packed, and can easily go in the wagons. When erected, they are secured by ropes and pegs in the ordinary way.

Parc d'aviation.—In the park there are three heavy mechanical transport wagons (*camions*) and one workshop wagon (*camion atelier*) for each *escadrille*. One officer, one intendant, and 20 rank and file, including drivers and mechanics, form the personnel of the park.

The *camions* are about 19 feet long on solid rubber tyred wheels. The driving wheels are in front and transmission is by chain; the wagon body is covered by a canvas tilt. The *camions* carry oil, motor spirit, a complete aeroplane engine, and spare parts. One of the wagons is fitted with a stand on which an aeroplane engine can be tested before being fitted to the aeroplane. The *camions*, like the *fourgons*, are fitted with a limber-hook for a trailer.

The workshop wagon (Crochat-Colardeau system) is 30 feet long. It has box sides which are hinged along the bottom edge; they are let down when the workshop is to be used, and are supported by three steel legs so as to form a platform for the workmen. The wagon contains a lathe and a band saw. Both machines are electrically driven, the power being generated by a dynamo driven off the main propelling engine (24 h.p., 4 cylinders , which can be disconnected from the road wheels and transmission gear. This dynamo also furnishes the power for five incandescent electric lights in the wagon, which enable work to be carried on at night.

The wagon contains in addition, anvils, vices and boxes of tools; all field repairs to the aeroplane and its motor can be carried out.

Reserve de ravitaillement.—This echelon is located on the railway line and only moves by rail. It is a depôt for all stores required for the upkeep of the aeroplanes and their transport.

The transport work between the *reserve* and the *escadrilles*, or the aviation park, is carried out by the mechanical transport wagons of the two latter formations.

Experimental wireless installation.—One Maurice Farman biplane was fitted with an experimental wireless installation, designed by M. Rouzet, who was present at the manœuvres as a corporal of the reserve. The maximum power is 250 watts, supplied by a dynamo driven off the main aeroplane engine. The apparatus is fixed at the back of the pilot's seat, a length of wire 100 metres long being run off a reel below the plane by means of a weight.

Messages have been transmitted up to 100 kilometres ($62\frac{1}{2}$ miles). So far, it has only been possible to transmit, the noise of the motor making it impossible to receive messages. An experiment is shortly to be tried with a receiving apparatus, consisting of an incandescent lamp made to glow by the received waves, the message being read by the alternate glowing and extinguishing of the lamp on the Morse system. It is thought that this system will hardly be suitable for long messages, but that the receipt of a transmitted message may at least be acknowledged.

Distinguishing marks.—The pilot wore an armlet showing to which side he belonged. This, however, was indistinguishable at 50 yards distance. Practically, it was impossible to tell to which side aeroplanes belonged, except perhaps for technical experts.

Hostile aeroplanes observed the rule of the road, but made no attempt to interfere with each other. An aeroplane was not forced to retire or descend if two or more hostile aircraft were encountered. Aircraft were not fired on, but were liable to capture if they landed in the enemy's country.

Armament.—It was stated that experiments were proceeding, and that the future arm was likely to be some form of automatic rifle rather than a machine gun, owing to the weight of the latter.

Supply and Transport.

Supply.

It is necessary first to explain the exact meaning of the term *éventuel*, which is so often used in connection with supply of various sorts, and the true significance of which does not appear to be understood.

1. The daily issue of meat is distinct from that of other supplies (unless preserved meat is issued).
2. In addition to this meat issue, the *Service de l'Arrière* send daily to points where regimental trains (*trains regimentaires*) refill, one complete day's supply for the army, of bread, oats and groceries. These daily consignments, which are made up according to the paper strength of the larger units, are called *ravitaillements quotidiens,* and are automatic, that is, they are standing daily consignments made without any demand having been put forward.

All issues or consignments other than the foregoing, that is to say, consignments of any sort issued or made in consequence of a special demand, are called *ravitaillements éventuels.*

Now, bearing in mind that nothing of what follows applies to fresh meat, each day the D.E.S. (*Directeur des Etapes et des Services*)* of the army arranges with the regulating station (*gare regulatrice*) for the despatch to suitable railway stations (according to the intentions of the commander-in-chief) of supply trains, containing bread, groceries and oats for all the component parts of the army. The D.E.S. is informed by the *direction des chemins de fer* or *commission regulatrice* at what time these trains may be expected at their respective destinations. Then, with this information in hand, Army Orders, Part II., are prepared. In these, general officers commanding army corps find instructions as to stations and times at which their respective commands can refill. In Part II. of Army Corps Orders, divisional commanders similarly find where they shall send for supplies for their divisions.

In the 9th Army Corps, as there was no mechanical transport available for bread, groceries and oats, the supply sections (*sections de ravitaillement*) of regimental trains were sent daily under their respective *officiers d'approvisionnement†* (regimental officer i/c supply and transport), to refill direct from the railway trucks.

---

* *See* page 12.
† In each regiment, &c., there is an *officier d'approvisionnement* who is in charge of supplies and transport. He is appointed in the same way as an adjutant in our service. His duties embrace all those of our regimental quartermaster, but he has far greater scope than the latter. It is his business to requisition or purchase everything which his regiment requires.

Where mechanical transport is available for divisional supply columns, the mechanical transport columns load at railway stations indicated to the commanding officers of mechanical transport units by the D.E.S. of the army, and proceed to rendezvous (*points de première destination*) detailed in Army and Army Corps Orders, Part II. At these rendezvous they are directed to refilling points (*points de contact*), where they meet the regimental trains and where loads are transferred to regimental vehicles and to regimental charge.

Supplies remain on charge of the accountant (*gestionnaire**) of the regulating station until they are actually handed over to the *officiers d'approvisionnement* of the regimental trains. If, for some reason or other, a unit fails to draw its supplies, or if, as is invariably the case, there is a surplus to requirements at the refilling points after the issue is completed, the "remains," go back to the regulating station whence they came.

In order to effect this satisfactorily, it is obvious that some efficient arrangement of personnel must be devised. The French system is as follows :—

Each railway train coming from the regulating station with supplies, brings its own personnel. For instance, in a supply train loaded with supplies for an army corps, there will be one *officier d'administration** per division, and one for the corps troops. In addition to these officers there will be a party of men (8 to 10) belonging to the *Manutention*,† and further, a fatigue party 12 to 15 strong, drawn from one of the regiments under the orders of the *Commandant d'étapes de la gare regulatrice*,* to assist in unloading the trucks. On arrival at the railway station, each of the *officiers d'administration* supervises the unloading of the railway trucks into the vehicles intended for the division or corps troops for which he is responsible.

If the regimental train are loading from the railway trucks direct, the issues are made then and there by the *officiers d'administration* to the regimental officers i/c supply and transport.

The supplies are loaded into regimental vehicles by regimental parties who have come with them (two men per wagon). On completion of the issue, the *officiers d'administration*, together with the fatigue parties return to the regulating station in the same train in which they came, taking with them any surplus supplies.

If, on the other hand, mechanical transport divisional supply columns load from the railway trucks, then the *officiers d'administration* and their parties who have come by rail not only unload the railway trucks, but also load up the supply column lorries, each *officier d'administration* being responsible for the loading of the lorries intended for his particular division, &c. The *officiers d'administration* and their men then take their places in the lorries and go in them to the rendezvous and on to the refilling points, where they meet the regimental train and issue to the *officiers d'approvisionnement* in exactly the same manner as described above in the case of the issue taking place at railway stations.

---

* The duties of these officers are given on pages 44 and 45.
† *See* page 44.

On completion of the issue at the refilling points, the *officiers d'administration* and their men, together with supplies remaining on hand, stay in the lorries until these reach a railway station to refill from the next supply train. With this train there of course comes another party of *officiers d'administration* and men. The latter party in their turn go out with the loaded lorries, whilst the officers and men of the first party return to the regulating station. It is not considered that by this system any wastage of transport occurs, as the lorries would have to go back to the railway in any case, and the empty railway trains would likewise have to return to the regulating station. Whether there is wastage of personnel is another matter.

It should be understood that the advanced base depot (*station magazin*) each day automatically sends forward to the regulating station in supply trains the quantity of supplies required for the *ravitaillement quotidien* of the entire army. These supply trains are loaded up in bulk. All issues other than the standing daily ones are *éventuel* issues. The latter provide for the supply (as far as the Army Service Corps is concerned) of biscuits, preserved meat, salt soup, live cattle, compressed hay and stores of that description.

Éventuel issues, as mentioned above, are only sent to the troops in accordance with special indents. The demands are made accurately, and no surplus is returned to regulating stations. The despatch and issue of *ravitaillements éventuels* are the same as in the case of *ravitaillements quotidiens*.

Fresh meat.—The system of supply of fresh meat in the field has recently been changed, and now is as follows:—A *parc de bétail de corps d'armée* is formed at some convenient centre, if possible on a railway. A municipal *abattoir* is generally selected for this purpose. Cattle are bought locally by the *sous intendant* in charge, assisted by his veterinary officer. If cattle cannot be bought locally or in the requisitioning zone under his authority, the *sous intendant* addresses an urgent demand, through the *intendant* of the army corps, to the D.E.S. of the army. The D.E.S. treats this demand as a *ravitaillement éventuel*, and forwards the cattle required from the regulating station or the advanced base depot.

The cattle are slaughtered daily in the *abattoir*, and are sent to units in mechanical transport lorries which are part of the establishment of the *parc de bétail* of the army corps and are in charge of a lieutenant of the transport corps (*train des équipages*).

Each day the general officer commanding army corps informs the *sous intendant* in charge *parc de bétail* of the time and place for the rendezvous for his lorries. The troops get this information from Part II., Army Corps Orders.

Throughout the manœuvres, the rendezvous and refilling points for meat were always different from those of other supplies. The *officiers d'administration* and working parties of the *parc de bétail* go to the rendezvous in the lorries and thence are directed to refilling points where they meet the regimental meat carts (*voitures à viande*) under the charge of the regimental *officiers d'approvisionnement*, to whom they make the issues. At these refilling points the *officiers d'approvisionnement* inform the *officiers d'administration* of the number of meat rations, &c., which they will respectively require for the next day.

On completion of the issue, the *officiers d'administration* and party return to the *parc de bétail* with the empty lorries.

When the troops move too far away from the *abattoir*, the *parc de bétail*, on the order of the general officer commanding army corps, moves on to a more convenient spot, recommended by the *intendant*. The stock of cattle on hand is always kept low, so that when a move becomes necessary, the two or three beasts remaining over are slaughtered and taken on in the lorries.

Other points in connection with the supply of fresh meat are as follows :—

The regimental supply and transport officer with each unit has on his charge the necessary tools, &c., for slaughtering, in case it should become necessary to live on the country.

In the *parc de bétail*, there is a most careful system of branding the cattle in order to prevent the exchange of animals. Not only is each animal branded on the near quarter, but a metal button bearing his number is clipped on to the near ear, in order to check the weight of the animal from the time of purchase alive until actual issue.

The cattle that were slaughtered compared favourably with those issued to our troops ; they were prize cattle purchased at 5*d.* per lb. live weight over weighbridge. They were slaughtered during the morning, cut up in quarters and placed in mechanical transport lorries which reached the regimental train at about 3.30 p.m. They were then placed in the meat wagons (*voitures à viande*) to be cooked and consumed on the evening of the following day.

The *voitures à viande,* of which there is one per battalion, form part of the 1st line transport, and go everywhere with the troops. They are small covered two-horse vans, with perforated zinc sides and back ; they have coarse canvas curtains outside to keep out the sun and dust, and are fitted with a ventilating roof. They are provided with steps which let down to the ground, and can carry 900 kilos. (1,984 lbs.).

The meat lorries are covered motor vans, painted red, with panels of perforated zinc at the front, sides and back. They have four steel rods running from the back to the front of the body just below the roof. On these rods are steel hooks for the meat. There are canvas curtains along the sides, and a hook on the framework of the door on which to hang scales. A lamp is fixed on the inside of the door so that when the door is open, the inside of the vehicle is illuminated. In each vehicle are lockers for butchers' implements.

Seven quarters can hang comfortably from each of the four rods. Each lorry can carry 2,000 to 2,500 kilos. (roughly 4,400 to 5,500 lbs.), the latter weight being the maximum load allowed.

The *parc de bétail* of each army corps (strength, about 20,500) on manœuvres, had eight such lorries (including three spare), and one lorry for tools, petrol, kits, &c. Each lorry is supposed to carry sufficient meat for one day for one brigade on the peace establishment, or for one regiment at war strength.

Petrol.—The supply of petrol, lubricating oil, &c., had not yet been systematized. Experiments have been carried out, but the results are private.

It is, however, certain that petrol will be sent to the front as a *ravitaillement éventuel* (*i.e.,* on demand through general officers commanding to D.E.S) from the regulating station, and will be issued in the same manner as other *éventuel* issues.

It appears that travelling tanks are not used. No petrol, &c., whether by rail or road, was seen in any special vehicle. The lubricating oil was in small cylinders, and the petrol was either in 50 litres (11-gallon) tins, fitted with screw caps and taps, or in boxes containing 10 or 12 half-gallon tins. No special pains appeared to be taken to store these tins; they were merely kept piled up in lorries of which one is told off to each section of the mechanical transport company for the purpose of carrying petrol, oil, kits and tools, &c.

Hay.—A force in the field is supplied with hay either from local resources or, as compressed hay, from supply depots. It is never carried with troops.

From local resources.—The *officier d'approvisionnement* is furnished with a list of prices which must not be exceeded. The contents of this list are secret, and it is his duty to make the most advantageous bargains possible. If necessary, he can (by signed requisitions) compel sales by inhabitants within the prices laid down in his list. He pays for his purchases in cash, and the vendor's receipts are the vouchers to his account. The quantities purchased by him must agree with the strength of his unit, and he is responsible for over-purchases when his account is finally balanced. He makes what arrangements he likes for collection, either sending some of the regimental vehicles to the vendor's premises, or arranging for delivery at his own cantonments by the vendor. The procedure is the same for straw bedding for the men, fuel, wood, &c. The system works admirably, purchases being made at the end of the day's march, either just before the troops reach their halting places, or immediately on their arrival there. Moreover, the procedure is so well known throughout France, that peasants are daily to be met driving loads of hay in the direction where they think the troops are likely to be, on the off chance of effecting a sale, a fair price always being obtainable and cash payments being very much appreciated.

Compressed hay (éventuel supply).—If from some cause or other hay is not obtainable by local purchase, *officiers d'approvisionnement* notify the *sous intendant* of their division of their requirements; a demand is put forward through the channels used for *ravitaillements éventuels*, and a daily supply of compressed hay is added to the *ravitaillement quotidien*, and is sent down daily by order of the D.E.S. from the regulating station or advanced base depot until local resources are available.

Bread.—Bread is supplied to a force in the field by *ravitaillement quotidien*. It is baked at the advanced base depot and is thence sent daily in bulk to the regulating station, where it is packed in railway trains which go forward daily to the *gares de ravitaillement*.

One point only is worthy of notice in this connection; for various reasons as stated before, supplies, including bread, are daily sent back to the *gare regulatrice* from the front, and it is a hard and fast rule that any *bread* thus sent back *must* go forward again in the supply railway trains the following day. It will thus be seen that at the regulating station there may occur a considerable accumulation of bread that has arrived from the advanced base depot, and has not been sent forward owing to " returns from the front " which have taken precedence. It is the business of the *commandant d'étapes de la gare regulatrice*, advised by the *sous intendant* on his staff, so to control the output of the advanced base depot that such accumulations shall not exceed a suitable working margin. It must also be remembered that a Frenchman is an enormous bread eater; his ration is 750 grammes (roughly 1½ lb.) per day. He generally breaks it up into his soup, so if it is a little stale, it does not matter very much. The bread supplied in the field (*pain biscuité*), however, is different from ordinary bread, and is still quite fit for consumption 8 days after baking.

No special vehicles are provided for the transport of bread. The careless way in which bread is handled is a source of great delay when transferring it from one vehicle into another, especially when loading lorries from the railway trucks, and when loading wagons of the regimental train from the mechanical transport lorries. It is proposed to pack loaves, by tens, in nets at the regulating stations, but the experiment contemplates the daily return of these nets, and this difficulty is proving an obstacle to the adoption of the scheme.

Neither field bakeries nor field butcheries were in use at these manœuvres.

Times taken to load, &c.—Supplies throughout the manœuvres were never transferred to regimental vehicles before the day's march began. When the regimental train refilled direct from the railway trucks at stations, the hours for doing so varied from 8.30 a.m. to 1.30 p.m.

The actual times taken by the mechanical transport supply column to load up supplies for the 11th Army Corps were as follows :—

12th September, began at 10.30 a.m., finished at 12.40 p.m.

13th September (ammunition), began at 11.10 a.m., finished at 11.40 a.m.

14th September, began at 10.20 a.m., finished at 12.40 p.m.

15th September, began at 10.10 a.m., finished at 11.45 a.m.

16th September, began at 10.30 a.m., finished at 12.16 p.m.

As a rule of thumb, the French allow 2 hours for transferring from vehicles to vehicles the supplies for one division. As a matter of fact, they generally took less than the time allowed.

It took an average of 16 minutes to load a regimental bread van with 1,000 rations, *i.e.*, 1,500 lbs.

Rations carried on the man.—At the end of the day's march, and before the evening meal, the French soldier has on him—

(*a.*) 2 days' *vivres de reserve.**
(*b.*) ½ ration of bread.
1 ration of dried vegetables and salt.

He at once draws from the *voiture à viande* his ration of fresh meat and lard, and cooks it; he consumes half of it and the whole of (*b*), keeping half the meat to eat cold at the next day's long halt. That evening he draws one ration of bread, dried vegetables, salt, sugar and coffee.

The position is therefore as follows:—

(*a.*) On the man $\left\{\begin{array}{l}\text{Half-a-day's ration.} \\ \text{1 day's } \textit{vivres de reserve.}\end{array}\right.$

(*b.*) In 1st Line transport (*train de combat*). $\left\{\begin{array}{l}\text{In } \textit{voiture à viande}, \text{ 1 day's meat and lard} \\ \text{for cooking.} \\ \text{In } \textit{voiture à vivres}, \text{ 1 day's } \textit{vivres de reserve.}\end{array}\right.$

(*c.*) In *trains regimentaires* (*2e. ligne.*) $\left\{\begin{array}{l}\text{2 days' } \textit{vivres de reserve} \text{ rations, of which} \\ \text{one ration includes preserved meat,} \\ \text{and the other, extra sugar but no meat.}\end{array}\right.$

Travelling kitchens (*cuisines roulantes*).—No travelling kitchens were seen during the manœuvres, and it was stated that it is proposed to do away with them. The reasons for this decision are that—

They increased the length of the regimental train, already too long.
In time of battle they were in the way.
When once left behind they did not catch up again when wanted.
If upset they were never righted.

Transport.

Inter-communication.—The means of inter-communication between the field army and the mechanical transport are primarily the telegraph service (field or otherwise) and motor cyclists; the D.E.S. can pick up the supply and transport columns daily at the following points :—

(*a.*) Railhead (*gare de ravitaillement*).
(*b.*) Rendezvous (*point de première destination*).
(*c.*) Refilling point (*point de contact*), by means of the staff *officier de ravitaillement* (*see* page 13).
(*d.*) Quarters at night.

The *ordre particulier* which the officer commanding mechanical transport unit receives each day from the D.E.S. contains as a rule instructions as to where he is to spend the night. If this information is not embodied in the *ordre particulier*, it is sent to him later by the D.E.S. at either (*a*), (*b*) or (*c*) (*see* above). Similarly, any modifications in his original orders are sent to him in the same way.

---

* He really only carries one day's *vivres de reserve* on his person, one day's *vivres de reserve* being carried for him in the *voiture à vivres*.

The channel of communication between a mechanical transport column and the headquarters and line of communication staffs, or in other words with the D.E.S., are—

1. For direct *ordres particuliers* from D.E.S. — Through the staff officer of the D.E.S. forwarded by wire or by motor cyclist.

2. For orders as to rendezvous. — These appear in Part II. of Army Orders, which in theory should reach the mechanical transport column at its quarters each night. They very seldom do, but as the *ordre particulier* from the D.E.S. contains this information, it does not matter whether they do or not.

3. For modification or cancellation of orders already issued.— As stated above, mechanical transport columns may be intercepted during the day at various points by direct messages.

With regard to channels of communication between mechanical transport columns and units. During the manœuvres motor cyclists met the mechanical transport column at rendezvous with written messages from the chiefs of the staffs of divisions, and the instructions from one or other of these divisions always included instructions as to the refilling points for corps troops.

It will thus be seen that in a French army the D.E.S. can place his hand at practically any time of the day on mechanical transport columns. The longest period that he can be out of touch with them is three hours; these three hours being those during which the mechanical transport column is actually moving from *gares de ravitaillement* to the rendezvous. This statement is arrived at as follows:—

Maximum distance of single journey from *gares de ravitaillement* = 65 kilometres.

Average distance from *points de première destination* to *points de contact* = 10 kilometres.

Therefore, average maximum distance from *gares de ravitaillement* to *points de première destination* = 55 kilometres.

Rate of progression of mechanical transport columns in actual practice = 18 kilometres per hour.

Extent to which mechanical transport is utilized.—The new system of mechanical transport supply columns was only tried in General Galliéni's Western (or Blue) Army. Incorporated with this army were two companies of mechanical transport, one heavy company (*Compagnie Lourde Automobile*), capable of coping with the requirements of an army corps (20,500 men, 3,000 horses), and one light company (*Compagnie Légère Automobile*), capable of supplying the wants of one cavalry division (2,500 men, 3,000 horses).

These mechanical transport units were not necessarily to be employed with any particular army corps; that is to say, the heavy company might convey supplies on one day for the 11th Army Corps, and on the next day for the 10th Army Corps. As a matter of fact it was employed with the 11th Army Corps throughout the manœuvres, whilst the *Compagnie Légère* acted throughout in a similar manner for the 1st Cavalry Division.

With regard to the utilization of mechanical transport in the future, it is probable that mechanical transport will, as soon as practicable, take the place of the following organizations throughout the army. The change will be effected gradually and probably in the following order (*see* also below in this connection):—

1. *Service de ravitaillement en vivres* (supplies).
2. *Service de ravitaillement en munitions* (ammunition).
3. *Service d'évacuations et de santé* (medical services).

The heavy company was divided into three sections. One section was told off daily to carry the supplies, &c., for the corps troops, and the other two for the 21st and 22nd Infantry Divisions.

In war, there will be four sections instead of three, as the corps troops will require two sections instead of one.

The sections were of unequal strength both as regards personnel and *matériel*. This was unavoidable owing to the lack of homogeneity in the lorries; the company was therefore divided into one section of eight vehicles and two sections of six vehicles each.

In each section, one lorry was told off to carry extra tools, kits, petrol, grease, &c. It travelled in rear of its section, and was a sort of first-aid workshop. The commanders of sections (in motor cars) travelled in rear of their respective sections. The commander of the mechanical transport column went on ahead in his car to reconnoitre refilling points. He kept touch on these occasions with his column by means of a motor cyclist. This system appeared capable of improvement. As it was, on more than one occasion, the leading lorries of sections failed to take the proper turnings on the road.

Method of transferring loads from rail to mechanical transport.— The supply railway train arrived loaded up with supplies for the *ravitaillement quotidien* of an army corps. On its arrival an officer alighted and informed all concerned that he belonged to the *commission des gares*. He was a sort of travelling railway transport officer and had come from the regulating station. He at once took charge of the station and traffic, but delegated his technical railway duties to the stationmaster. He posted sentries (from a party of men who had come with him) at the entrances and exits of the station. He then ordered the stationmaster to shunt the supply trucks into the desired positions. As soon as these were ready, he reported accordingly to the officer commanding mechanical transport column, and the 1st section of the supply column came and loaded up for one division, under the supervision of the *officier d'administration* responsible for that division.

The loading. &c., as before described, was done by the men of the parties who had, together with the *officier d'administration*, arrived in the supply train. Six railway trucks were worked at one time, the lorries backing on to the side of the trucks and at right angles to them.

As soon as one section had loaded up it was moved off by the *officier de la commission des gares*, and formed up in column of route, under its own officer, outside the station yard, on the road, where it waited for the rest of the column. Great delay always occurred in opening boxes containing coffee, sugar, &c., and in handling bread.

The *officier de la commission des gares* returned to the regulating station in the empty supply train.

No system of mechanical transport for ammunition supply has yet been evolved.

When the War Office hires lorries, the specification lays down that they must be capable of transporting supplies, *ammunition* and sick. It is therefore quite certain that ammunition supply is contemplated, but the general idea in France is that mechanical transport units may be used for anything. They are *organes d'armée*, and one day may be doing supply work and on the next day something else, at the discretion of the supreme commander. Thus the *Compagnie Lourde Automobile* was taken away from the supply service on the 13th September, and was employed on ammunition service, while on the 15th September, lorries were used to convey two companies of infantry; 22 men travelled on each lorry sitting on their packs.

Subsidy scheme.—The subsidy scheme at present in force for the requisitioning of mechanically propelled vehicles is in the main as follows :—

A central committee (*commission centrale*) composed of technical experts, and also of representatives of the leading motor manufacturing firms meets each year at Versailles. Various new types of lorries are brought before this committee by their manufacturers. They embody the requirements of the War Department which during the year have been notified to the manufacturers and are based on the requirements of the service and on the industrial improvements of the last 12 months. The vehicles are subjected to the most searching tests of endurance for 8 consecutive days, during which their component parts are stamped to safeguard the committee against fraudulent exchanges. At the conclusion of the tests, the lorries which are found to be of the most suitable type are awarded a Government premium.

This premium does not go to the manufacturer. The War Minister found that if a premium was granted to the manufacturers, the latter merely turned out good vehicles which they sold to foreign Powers, whereas it was desired to create a reserve of suitable privately owned lorries on which the War Department would have first call. The following scheme was, therefore, devised :—

The premiums, on a sliding scale (according to power of vehicle) go to the purchaser, and benefit the manufacturer indirectly, in that they increase the sales of his lorries by cheapening them to the purchaser, who receives monetary assistance from the Government when buying the right type of vehicle. Thus the highest premium is 3,500 frs. (140*l.*) to the purchaser on purchase, and 40*l.* a-year for upkeep, for the three succeeding years. A good lorry ordinarily costing 880*l.*, in this way really only costs the purchaser 740*l.* The Government has a call on it for 3 years, and by the end of that time does not require it any more, as by then it is out of date.

Individuals desiring to buy a lorry and to benefit by the premium, must apply in writing to the general officer commanding army corps of the district in which they are domiciled. These applications are forwarded to the district committee (*commission regionale*) which is composed very much in the same way as the central committee, excepting that manufacturing firms are not represented on it, and that it includes the local mayor who watches over the interests of the purchasers. Applicants to purchase a *type primé* enjoy priority over those who apply for assistance in the purchase of any other type, and as the credits voted for these premiums are limited and the applications numerous, in practice only purchasers of *type primés* receive Government assistance.

During the subsidized period (three years) lorries must not cross the frontier for more than 15 days. They must bear a metal plate in a certain prescribed position, and they are subject to inspection twice a year (when their half-yearly upkeep premium is paid to their owners), and at such other time as may be ordered. If they change hands, the new owner must be bound by the same rules as the original purchaser, and the *commission regionale* must be notified of the change.

In addition to this, there is an annual census of motor vehicles. ALL motor vehicles in France are kept registered at the nearest *mairie*. An extract is made each year from this register and a return of the *poids lourds* lorries (8 seats, or capable of carrying 1,650 lbs.) is rendered to the headquarters of the army corps of the district. This return gives a description of each vehicle, and records the military obligations of the driver.

Each army corps throughout France recapitulates its own district returns into *tableau de classement*, which are placed in the hands of local *commissions de classement*, which advertise in all the communes the fact that they will sit on certain days at certain centres for compulsory inspection of motor vehicles. The *commission de classement* also has to check annually its *tableau de classement* with the registers of all the *mairies* with which it deals. All the mechanically propelled vehicles on the *tableau de classement* are brought before the commission by their properly accredited drivers, whose military status is checked then and there by the local recruiting officer. The vehicles themselves are classed according to condition, and the suitable ones are entered in a *tableau de requisition*. These make up a set of complete returns of every motor in France fit to be requisitioned, with the name and military service obligations of its actual driver, who must personally present himself with it when called up on mobilization. The vehicle is then purchased and taken over on the spot.

Such is the rough outline of the system in force. Owners of cars gladly submit to the regulations, as they thus gain the advantage of doing their reserve service as drivers of their own cars, instead of having to march in regiments. The subsidy scheme appears to be a great success, and, in the 9th Army Corps alone, the applicants in the last 12 months for subsidized purchases have increased from 63 to 189.

Fifteen metres is laid down as the minimum distance between vehicles when travelling on the road, and 50 metres as the maximum. They generally try to travel just clear of one another's dust.

The maximum distance of the daily round trip for mechanical transport is laid down as 130 kilometres (80¾ miles).

No mechanical transport vehicles were fitted or used for the conveyance of sick during the manœuvres, and although experiments had been made in the past by fitting stretchers to uprights, no vehicles have so far been fitted finally for this purpose.

The question of hours of work and rest for the personnel of mechanical transport does not appear to have been considered at all. It is thought that the fact of there being two drivers to each lorry removes any cause for anxiety as far as rest for either of the drivers is concerned. Further, it is considered, in view of the system in force in France, by which mechanical transport only begins loading quite late in the morning, that the mechanical transport personnel will be much better off as regards hours of rest than any other branch of the army.

Service de l'intendance.—The *Service de l'intendance* is divided into two broad divisions of officers—

    (*a.*) *Intendants* (including *sous intendants*).

    (*b.*) *Officiers d'administration.*

(*a.*) The *intendants* perform administrative duties and are responsible that the administrative orders of the command are carried out. They are generally officers drawn from the various arms of the service who apply to go to the *Intendance*. They are put through a severe course of instruction, and after a period of probation as *adjoints*, they become *sous intendants*. The lowest rank in this branch of the *Intendance* is that of *commandant* (equivalent to our major). They receive extra pay and their advancement is comparatively very rapid. Although they wear the four stripes of *commandant* they are never called by their rank, but always *Monsieur l'Intendant* or *Monsieur le Sous Intendant*, and although they exercise command over their own personnel, they cannot ordinarily go beyond that.

(*b.*) The *officiers d'administration* are quite a different organization; their work consists of the actual executive performance of those duties which appertain to administrative services. *Officiers d'administration* are invariably promoted from the ranks; it is a *sine qua non* that they should have been non-commissioned officers well reported on in the various arms of the service. They are sent through a very technical course of instruction at Vincennes, where are the headquarters of the *Manutention Militaire* (military bakery, &c.), and they can never rise above the rank of major.

A *gestionnaire* is an officer of the *Service de l'intendance* accountable for some " charge " or other. He may be in charge of supply stores or stores of any kind.

All *gestionnaires* are compelled to lodge with the Treasury actual cash to the value of the material of which they have charge. For instance, the *gestionnaire* of one store had lodged 6,000*l.* with the *trésor* in Paris. Government allows *gestionnaires* interest at the rate of 8 per cent. per annum on the amounts thus lodged with the Treasury as security, and this high rate of interest enables officers appointed *gestionnaires* to borrow without much difficulty the necessary sums which they have to lodge.

Commandant d'étapes de la gare regulatrice.—This officer, who holds a very important appointment, is generally a senior staff officer.

He is the military commandant of the regulating station, and as such exercises direct authority over the troops stationed in the vicinity of that station. He is responsible to the D.E.S., and is assisted by a large staff of officers, which includes not only officers of the *Intendance* and *Administration* (12 to 14 in number), but also other officers representing the *Services de l'Armée*. He will thus have on his staff an officer each of the artillery, engineers, medical service, provost marshal's department, pay and postal department, besides a large staff of clerks and labourers.

The officers of this staff act as his advisers on the technicalities of their respective departments. He also disposes of officers (generally belonging to the reserve) who are on the *commission des gares*. He sends these officers up and down the line as he (advised by the *commission des gares*) thinks fit, to organize, as may be necessary, the policing or management of railway stations to which supply, ammunition or store trains may be despatched. To these railway stations he also sends such detachments of men or fatigue parties from the troops under his command as the *commission des gares* may consider necessary.

He is responsible for stores in his command and their adequate upkeep; for the despatch of correctly constituted supply trains to the appointed stations at the front, and for the regular arrival of bulk supply trains from the *station magazin*. He is responsible for " quantities " of *ravitaillements quotidiens*, and the proper despatch of *ravitaillements éventuels*. With regard to all these responsibilities he works *with* the *commission regulatrice*, which runs the train service, but cannot under any circumstances give an *order* to the *commission regulatrice*, which is under the *directeur des chemins de fer*.

Billets.

The French believe that the worst billet is better than the best bivouac. They ascribe half the evils to which horeflesh is due, such as lack of condition, sore backs, &c., to exposure in bad weather.

During the cavalry manœuvres (*see* pages 8 and 9) the three divisions were billeted in different parts of the training area. The law forbids the billeting of troops in one place for more than three nights in succession, so after three days, all the divisions moved on to fresh billeting areas. The 2nd Division, after two days' absence from the place where it was first located, returned to its old quarters.

The distribution of troops in billets seemed to present no difficulties. No elaborate preliminary arrangements were made. On the morning of the change of quarters, billeting officers went on straight to the new area and, in consultation with the mayors, arranged for the distribution of the troops. An hour or two before the troops finished their work, regimental billeting parties

were sent ahead to divide up the regimental area, so that when
the troops arrived they found all the doors marked in chalk with
the designation of the unit and the number of men each house
was to hold, viz. :—

> 31 D to mean 31st Dragoons.
> 2 E   „   2nd Squadron.
> 3 P   „   3rd Troop.
> 5 C   „   5 horses.

In most places in Europe, the farms are situated in the villages,
which in consequence are more suitable for the billeting of
mounted troops than in England. Generally speaking, each village
held as many horses as there were inhabitants.

It did not seem possible that the villages could hold more
horses, for every available stable, barn and shed seemed to be used
to its fullest capacity ; as long as there was a roof over the horse
it did not seem to matter how dark and ill-ventilated the stable,
or how rough and exposed the shed. Approximately, 1 yard was
allowed for each horse, so it seemed likely that many of the horses
would not lie down at all at night. When wooden poles, beams
or planks were available, improvised bails were erected between
the horses, one end of each piece of wood resting high up against
the wall in front of the horse's head and the other end standing
on the ground in rear of the horse's quarters. In this manner the
chance of injury to the horses by kicks was materially lessened.
Two other methods of avoiding kick injuries were noticed. To
each troop were given four kicking hobbles (*entraves*), which
consisted of two pieces of numnah, about 9 inches deep (one to be
fastened round each hind leg above the hock), and connected by
a rope or strap made to fasten at any length, but usually about
18 inches long. These were for the worst kickers. The other
method, the effectiveness of which was apparently not doubted,
was to attach an ordinary stirrup-iron to one hind leg of the horse
just above the fetlock.

As there were usually no rings in the walls to which to fasten
the horse's headropes, the forage ropes, or ropes or chains obtained
locally, were run along the wall, at a height of about 3 feet from
the ground. There were no heel pegs or other methods of tieing
up the horse's hind legs, except the shackle mentioned above.
The men sleep in out-houses, close to their horses, or in hay lofts
above them, and are welcome so long as their visits to the same
village are not too frequent.

Miscellaneous.

The Reserve Division.

Composition.—Perhaps the most interesting feature of the
manœuvres was the incorporation of the 54th (Reserve) Division.
This was composed of six reserve infantry regiments, four
from the 4th Corps (many of whom were Parisians), one from the
12th Corps (Angoulême), one from the 18th Corps (Saintes).
Attached to it were two (active) brigades of field artillery, one
(active) engineer company, and one (reserve) regiment of cavalry.

The divisional staff was constituted as it would be in war, with the addition of two attached staff college students.

The composition of the reserve regiments exactly corresponded to their war organization. Regiments were commanded by the lieutenant-colonel of their active regiment; battalions were commanded by reserve or retired officers; companies were commanded by lieutenants on the active list from the *cadre complementaire*; 24 non-commissioned officers per regiment were from the active regiment, the remainder were reservists; the men were all from the seven or eight senior classes of the reserve; that is to say, men between 25 and 32 years of age.

Work.—The division was concentrated at Ruchard, where it remained for three days. On the fourth day it marched 15½ miles to the site of its entrenched position on the Vienne, and dug itself in the next day. On the following three days it was employed in active operations.

The division as a whole had not much marching to do; certain battalions, however, had to work hard. One of these marched 22½ miles in 24 hours.

Remarks.—The result of this experiment was, on the whole, satisfactory. It is evident that the material is excellent, although certain weak points were disclosed in the organization.

The companies which averaged 150 men at Ruchard, averaged 135 at the end of manœuvres.

The weak points are said to be the following:—

1. The battalion commanders, all retired or reserve officers, were not up to the mark, and should be replaced by officers from the active list.

2. The lieutenants commanding companies were not all of the right type to deal with the older men forming the bulk of the battalions.

3. The reservist non-commissioned officers had, many of them, lost touch with military matters. The number of active non-commissioned officers should be increased.

4. The divisional and brigade commanders should, *as a rule,* be officers on the active list.

Points one, two and three have been dealt with since manœuvres in the new *Loi des Cadres.*

One point that was noticed with regard to the reserve division, was the perfunctory way it carried out its protective duties.

On 15th September, the very important bridge at Pont de Piles was only occupied by a telegraph detachment. This detachment was easily captured by the first patrol of Blue cavalry who appeared. The Blue cavalry never finally relinquished their hold on this bridge, which formed a most important factor in the next day's operations.

Attached officers.

The system of attaching officers (*stagiaires*) to arms other than their own is treated seriously. The officers so attached actually command units. Infantry officers were seen commanding squadrons, artillery officers commanding companies, &c. It was not

seen, however, whether attached officers of infantry or cavalry
actually commanded batteries or sections of artillery.

Students of the staff college of various arms were seen with
units in cavalry divisions and with the staff. In both cases they were
given cavalry work to perform; when attached to squadrons they
were given executive command of troops, both in billets and in
the field.

Weather.

A short but interesting discussion on weather took place at
the cavalry manœuvres. On each day it was fine from about
3 a.m. to 7 a.m., and an officer contended that such was generally
the case, and that dawn was the best time for reconnaissance,
troops then beginning to stir in their billets. The very early
mornings, too, were generally propitious for aircraft.

It was held that Manchurian experiences prove that, however fine the weather, heavy cannonades do not fail to bring
down torrential rain about the third day, and that in a big war
rain and Napoleon's fifth element—mud—will be the normal
conditions that troops must be prepared to face.

GERMANY.

The Imperial Manœuvres took place in Saxony, north of Dresden, on the 11th, 12th and 13th September. On the 9th and 10th September, cavalry exercises were carried out in connection with and leading up to the subsequent operations.

The opposing forces were organized as follows:—

Red Force.—Commander—General of Infantry von Bulow.

III. and XII. (1st Royal Saxon) Army Corps, including 2 cavalry divisions.

Blue Force.—Commander—Field-Marshal Freiherr von Hausen.

IV. and XIX. (2nd Royal Saxon) Army Corps, 3rd Cavalry Division and a Bavarian division of cavalry.

Total strength—110½ battalions, 129 squadrons, 128 batteries of horse and field artillery, 16 heavy batteries, 18 machine gun companies, 4 machine gun batteries (with cavalry divisions), 5 pioneer battalions, 4 cavalry pioneer detachments, 6 telephone detachments and 3 telephone sections, 5 divisional and 2 cavalry bridging trains, 2 wireless detachments, 4 signal companies, 4 aeroplane detachments, and 2 airships.

Country.—The theatre of operations was situated approximately in the triangle Dresden, Berlin, Leipzig. The country on either side of the Elbe is flat, but further to the west it is more broken and undulating. The ground is very highly cultivated. There are no fences, and troops can move almost anywhere off the roads. The villages consist of strongly built brick houses, and the numerous barns are remarkable for their size and solid construction. The River Elbe, which flows diagonally across the north-eastern corner of the manœuvre area, is from 250 to 300 yards wide in this section of its course. The stream is not swift and runs between steep banks of varying height.

Nature of the operations.—As in 1911, this year's manœuvres were designed to illustrate a possible campaign against a combination of European Powers.

Whilst the Blue, or home armies, were engaged in a struggle with a western Power, a Red or eastern army had crossed the Blue frontier and was advancing through Northern Bohemia and Neumark. After a decisive victory over the western Power, three Blue armies (Fourth, Fifth and First) were brought back by rail, and on the 8th September were approximately on the line Magdeburg—Jena, drawn up along the left bank of the River Saale. The three armies were disposed from north to south in the order given above. Blue reserve troops were posted along the Elbe and Erz Mountains. The First and Second Red Armies had on the same date reached the line Berlin—Dresden, and the Third Army was due south of them beyond the Erz Mountains. The Second Army was south of the First. In each case the central army (the Fifth Blue and the Second Red) was represented by the troops actually engaged; the other armies on the flanks were imaginary.

On the 9th September, the Second Red and the Fifth Blue Armies, conforming to the movements of the imaginary troops on their flanks, marched towards each other, preceded in each case by an independent cavalry corps. During the course of the day, the Red cavalry corps drove in the weak Blue reserve troops on the Elbe, and succeeded in establishing itself on the further bank of the stream before nightfall.

On the 10th September an engagement took place between the two hostile cavalry corps at Gaumnitz-Hugel. The infantry who were present with the Red cavalry and the faulty tactics of the Blue cavalry were responsible for the complete defeat and retirement of the latter. The Red troops did not pursue. On the 11th September the whole of the Second Red Army, covered by its cavalry, crossed the Elbe. On the 12th September a general engagement took place between the two opposing armies. The Blue commander was the first to move, and made his main attack against the Red right flank. After a hotly contested fight, both flanks of the Blue army were driven in, and the Red army was left in possession of the field. On the 13th September, the Blue army was just able to hold its own on its left flank, but its right or southern flank was severely defeated and driven back. Before the Red army could develop the success it had gained, the manœuvres were broken off.

The Passage of the Elbe River.

It may perhaps be said that the main feature of this year's Imperial Manœuvres was the passage of the Elbe by the Red army. Crossings were made at three points. Nothing had to be improvised, the three bridges being constructed with the equipment in possession of the troops, and as no serious opposition was offered by the enemy, there were no special difficulties. The work was on the whole well and quickly done.

(a.) *Passage of the Red cavalry corps.*—This was not seen by any British officers, but a credible eye-witness has stated that the Red cavalry approached the river without sending any patrols to the front, and made no pretence of reconnoitring the river or of finding the most suitable places for crossing. These were apparently known and had been determined beforehand, and the troops were marched straight to these points without further ado. The crossing was made in the usual manner by means of the steel pontoons furnished by the cavalry bridging trains. The horses (four on each side of a pontoon) were swum across, drawing the pontoons with them; they had on their bridles which were held by the men in the boats, which also contained the saddlery. The guns and wagons were ferried across by means of the usual rafts.

(b.) *Passage of the main body.*—The passage of the leading infantry was on the whole rapidly and skilfully made. Advantage was taken of any available cover and bends of the river to start unseen from the shore, and the nine or ten boats which suddenly pushed out into the stream from different points must have made accurate fire from the opposite bank difficult.

Twelve men carried each half-pontoon with ease; 22 men made up the approximate load for each whole boat. When crossing without horses, four men row, and one steers with an oar in the stern.

The bridges constructed were of the normal type (*see* Report on Foreign Manœuvres in 1910, pp. 155, 156 and 159) and do not, therefore, require detailed description here. The half-pontoons were joined together on the bank up stream and then floated down into their positions. The bridges built were chiefly by the pioneers, but infantry working parties also were detailed to assist. Horses and men seemed thoroughly accustomed to the work, and every man apparently knew what was expected of him. However, there was a shrewd idea prevalent in the III. and XII. Army Corps many months ago that the Elbe would be crossed during the Imperial Manœuvres. The passage of rivers was, therefore, practised very often during the previous summer.

Railway arrangements after the manœuvres.—It is believed that the railway arrangements after the manœuvres were very successful, and were carried out without any appreciable dislocation of the ordinary service of trains. It appears from newspaper reports that 163 trains were used to transport some 85,000 men, 1,800 horses, 1,500 vehicles and quantities of baggage and warlike stores. The trains were often very long, consisting of 40 to 50 carriages; 50 carriages were apparently the maximum number allowed in the case of any one train.

Direction of Manœuvres.

The system of control exercised by the directors of manœuvres in the initial disposition of the troops and through the medium of imaginary armies on either flanks of the troops engaged, was not different to other years (*see* Report on Foreign Manœuvres in 1911, pp. 22 and 23).

The movements of these imaginary armies were scrupulously worked out during the entire manœuvre period, and were made known to the troops by means of official maps and summaries of the situation, which were issued once or more daily. Maps and summaries dealing solely with the dispositions of the imaginary armies were also published.

The action of commanders was, therefore, as usual somewhat cramped, the strategical aspect of the operations having already been determined. Turning movements on a large scale and similar enterprises were also entirely out of the question. There is indeed evidence indicating that a kind of programme of events had been arranged beforehand, viz.: first day, strategical reconnaissance; second day, cavalry combat; third day, passage of the Elbe by the Red army; fourth and fifth days, general engagement between the two armies. To such lengths was this system of control carried. Unreal situations were not only permitted, but possibly insisted upon in order that the proper sequence of events or *tableaux* should not be upset.

The Imperial Manœuvres were continuous. During corps manœuvres, after each day's operations an outpost line was held for two or three hours. When it had been inspected, the troops went to their billets, reoccupying the line before operations began on the following day. At corps manœuvres, the commanders of sides were changed daily.

It was foreseen by the manœuvre section of the Great General Staff that if the two Prussian army corps detailed for this year's Imperial Manœuvres had been pitted against the two Saxon corps, a certain amount of undesirable feeling might have arisen. It was, therefore, arranged that one Prussian and one Saxon corps should fight on each side. Similarly, the Prussian commander of the Red side was furnished with a Saxon chief of the staff and the Saxon commander of the Blue side with a Prussian chief of the staff.

With so much that was admirable on the German Imperial Manœuvres, there was one noticeable feature by no means excellent. The troops did not " play the game." For instance, hostile patrols met in the most casual manner and, if no officers were present, supplied each other with information. Instances, showing like these, that the troops could not or would not throw themselves into the spirit of the operations, were constantly occurring.

Umpiring.—The system of umpiring was the same as that described in the " Report on Foreign Manœuvres in 1911," page 25.

The Emperor as Umpire-in-Chief put troops out of action on an unusually large scale. The whole of one (Blue) cavalry division was put out of action on the 10th September (*see* page 54). Entire brigades of infantry were also put out of action at a time.

One of the duties of the umpires is to keep the Directing Staff informed of every decision and movement made in the theatre of operations. The bulk of this work actually falls on the *Nachrichten-offiziere*, whose sole duty it is. The work was performed in a remarkably efficient manner. At Mudeln, where the headquarters of the Directing Staff were established in a large building, the most perfect order and system prevailed. The foreign attachés called on several occasions at these headquarters to obtain news of the situation, and information invariably was forthcoming as to the progress of events in every part of the theatre of operations up to the last quarter of an hour. Though the officers employed in this somewhat harassing work had only very few hours of sleep during the manœuvres, they never seemed overworked or unduly pressed for time ; while the admirable summaries of events and maps showing the situations point to their efficiency. Of course the elaborate preparations made before the manœuvres, the finger posts showing where various headquarters, &c., were situated, and the network of neutral telephones, made the work easier.

Casualties.—Casualty flags were carried, but were not used to any greater extent than in former years.

Officers and men who had been out of action removed their helmet covers, but did not this year remain in the firing line or with their units, as was the case last year. The novel arrangement tried on that occasion ("*see* Report on Foreign Manœuvres in 1911," page 27), apparently was considered a failure, as it was not repeated this year.

Maps.—Maps were issued to the troops on a generous scale; each battalion received 800.

Infantry.

It is difficult to say anything new about the Prussian and Saxon infantry. The spirit, discipline, endurance and marching powers of these men are in every sense admirable. If there is a weakness it is the leading of company officers. This weakness is most evident in the control of fire which is elementary and inefficient, as has already been reported. (*See* "Report on Foreign Manœuvres in 1911," page 42.) The company officers are also too much disposed to look to the rear for inspiration, not realizing that in the firing line they themselves are generally the best judges of what action is demanded by the situation. It is not desired to labour this last point too much; the many good qualities of German infantry officers are well known, but an increased measure of initiative would add greatly to their fighting value.

Attack.—The various deployments for attack were seldom made under cover, even if available. The change from column to line frequently took place within view and effective rifle range of the defence, and there was usually no difficulty in determining, from within the defensive line, the position of the flanks of the attack.

Forward movements in the attack were made with but few pauses, so that they could seldom be supported by infantry covering fire.

The rigid discipline in companies and sections appeared to leave no latitude for wide extension even when the ground permitted it. For example, in quite open country, there were frequently large areas of unoccupied ground between battalions or companies.

Defence.—When meeting an assault the infantry of the defence invariably left their trenches when the attackers approached to within about 150 yards, and met them with a counter charge just at the moment when rifle fire would have been highly effective.

Machine guns.

Machine guns appeared to play a somewhat minor rôle this year. The proportion of these guns to infantry is at present hardly large enough to make them very conspicuous. Six machine guns to 4,200 men, which was the strength of an infantry brigade on manœuvres, or one to each battalion, are hardly an adequate number. On the evening of the 11th September the Blue army was extended over a front of 30 kilometres (nearly 19 miles), and along the whole of that front there were only 8 machine gun companies. This state of affairs will be remedied in time.

No new ideas as to the employment of these weapons appear to be prevalent, and the previous remarks on this subject still hold good. (*See* " Report on Foreign Manœuvres in 1911," page 63.)

Cavalry.

Last year the independent cavalry were given as much opportunity of fighting against infantry and artillery as possible. This year strategical reconnaissance and a cavalry engagement on a large scale were the objects in view.

Before the cavalry engagement on the 10th September began, the opposing forces were drawn up on either side of a valley. The Blue cavalry commander elected to attack. The Red cavalry dismounted and, supported by artillery, fired at the enemy as he advanced, mounted, across the open. A brigade on the left of the Blue attack was unexpectedly met by the fire of one of the rifle battalions allotted to the Red cavalry. The fire of this battalion was judged to have annihilated the cavalry brigade. As was to be expected, the Blue cavalry were entirely defeated; half of the corps was put out of action, and it was ordered to retreat 20 kilometres ($12\frac{1}{2}$ miles) by the Umpire-in-Chief.

The most interesting feature of this engagement was the presence, with the cavalry division, of infantry who were conveyed by means of mechanical transport wagons, which were of the usual pattern for carrying supplies. A German officer explained the attachment of infantry to the independent cavalry as an attempt to reply to the cyclists accompanying the French cavalry.

Experimental method of carrying the carbine.—In some Saxon regiments the carbine was carried on the near side of the horse, attached to the wallet and to the saddle near the knee of the trooper. This method did not appear to possess any special advantages.

Artillery.

Very little was seen of the work of the artillery, but the tendency of this arm to support the infantry at all costs was, if possible, more pronounced than ever. On two occasions during the battle which raged round Oschatz on the 12th September, batteries were seen to sacrifice themselves in order to give all possible support to the infantry. The guns, which were laying direct, were worked until the hostile infantry came up and took possession of them.

One Krupp and one Ehrhardt gun for use against aircraft were again present this year. Light-ball pistols (*Leuchtpistolen*) were used to show that the guns were in action.

Signal Service.

Of the four existing Prussian telegraph battalions the whole of the 1st and 2nd and a portion of the 3rd and 4th Battalions took part in the Imperial Manœuvres.

A large number of reservists were called up to join the telegraph troops, including 104 men for 42 days, who were detailed for duty with the telephone detachments accompanying the infantry.

The telegraph troops were supplied with climbing irons, and made considerable use of them; the men concerned were very expert.

Air Service.

Two dirigible airships, one of the Zeppelin and one of the Parseval type (Z. III. and P. III.), and 38 aeroplanes took part in the Imperial Manœuvres.

Dirigibles.—Z. II. had also been detailed for duty, but was damaged early in September and therefore unable to be present.

P. III. was housed at Schenkendobern in one of the movable canvas hangars.

Very little is known of the actual work performed by these dirigibles, but the Emperor congratulated Count Zeppelin on the very accurate information which had been supplied by Z. III. Dirigibles were more conspicuous than aeroplanes, and appeared to be omnipresent and under complete control to move rapidly in any desired direction. Z. III., it is believed, always flew at a height of at least 6,000 feet.

Aeroplanes.—The official summaries did not this year comment on the excellence of the intelligence supplied by the aeroplanes, which were less conspicuous this year than last year, though nearly five times as many were present. The weather was not so favourable as in the 1911 manœuvres, but not bad enough to stop flying in any way.

The following took part:—

> 17 E. Rumpler Taube monoplanes.
> 17 Albatross biplanes.
> 2 Luftverkehr-Gesellschaft biplanes.
> 1 Wright machine.
> 1 Dorner monoplane.

Aeroplanes were housed in the movable canvas hangars mentioned above.

German military circles consider, it is said, that the chief lesson of the manœuvres with regard to aeroplanes was that the best results can be achieved with slow machines. The E. Rumpler Taube and Albatross biplanes were praised specially in this connection.

Supply and Transport.

Supply under service conditions, when no food is obtainable in the theatre of operations, was, for purposes of economy, only practised by the III. Army Corps and the cavalry.

Travelling kitchens were the first line of supply, and gave the usual satisfaction They filled up from regimental transport wagons, and these latter from supply trains.

The cavalry was supplied by means of mechanical transport.

Two-thirds of the mechanical transport battalion were present, but none of the three Prussian railway regiments, except a detachment of the 1st Railway Regiment, which was detailed to build the hangar for P. III. at Schenkendobern.

Miscellaneous.

Volunteer Motor Corps.—50 members of the German Volunteer Motor Corps placed themselves and their cars at the disposal of the military authorities for the manœuvres. 34 motor cyclists of the German Universal Automobile Club also took part in the operations, chiefly as orderlies.

The Emperor's headquarters.—The Emperor's headquarters which were established during the manœuvres at Limbach, consisted of four huts of wood and asbestos, set up in a clearing in a wood. The huts were all about 33 feet long and 13 feet wide, and could, it is said, in the field be set up by 60 pioneers in one day.

Motor boats.—The "Motor-Yacht Klub von Deutschland" provided some motor launches which were fitted with machine guns and searchlights. They were allotted to the Blue side and were used to oppose the crossing of the Elbe by the Red army. It is stated in the Press that, under the eyes of the Emperor, two motor boats succeeded in throwing into confusion the attempted crossing of two divisions of Red cavalry by surprise machine gun fire. These boats were, however, soon put out of action by the umpires.

GREECE.

The manœuvres took place between the 23rd May and the 2nd June, but military attachés were only invited to attend between the 29th May and the 2nd June.

The troops taking part consisted of the 1st and 2nd Divisions, and 5 additional infantry battalions. About 18,000 men were engaged, organized as follows :—

7 regiments and 2 battalions of infantry.
2 regiments of cavalry.
2 field artillery regiments, and
1 engineer regiment.

The manœuvres were a series of field days rather than a continuous operation, and the composition of the opposing sides was subject to change.

Infantry regiments were made up of 3 battalions, each of 4 companies, and the strength of companies varied between 125 and 170. A cavalry regiment had 4 squadrons with 80 horses. A field artillery regiment consisted of 2-3 groups (each of 2 batteries) and each battery had 4 guns, and 5 or 6 wagons.

In order to bring units up to the required strength, the whole of the 1906 class, and some untrained men of the 1903 class, were called out from the reserve.

Transport of troops by sea.—There is but little facility for transporting troops by rail in Greece. At the conclusion of the manœuvres a portion of the 1st Division, about 2,300 in all, were therefore sent home by sea. Seven transports were used for the purpose, none of which were of more than about 1,200 tons. They were not fitted up in any way. It was not possible to get any accurate information regarding the number of men per ton, but in one boat, estimated at about 1,000 tons, there were 690 men. The men in this ship landed at Volo at night with the assistance of the searchlights from two cruisers and by means of man-of-war boats, holding about 60 men each, and towed by launches. The horses were packed close together in the hold without any fittings. This was a risk, but fortunately the sea remained calm. The duration of the voyage was 24 hours.

Direction of Manœuvres.

The Director of the manœuvres was General Eydoux, the principal French reorganizing officer. Operations began at an hour which was fixed daily in orders and varied between 3 a.m. and 7 a.m. They ceased about noon and there were no night operations.

Umpiring.—The members of the French mission, each of them accompanied by a Greek officer, did the work of umpiring. Umpires were definitely allotted to the different arms, but they

had the power of giving decisions regarding arms other than their
own on either side if required to do so.   Their orders were :—

    1. To send back or temporarily put out of action any troops
        who were in a tactically impossible situation.
    2. To prevent the too rapid progress of an operation.
    3. To prevent opposing troops from approaching within
        100 yards of one another.
    4. To give an account of the course of the operations to the
        Director at the end of the day.

Infantry.

The Greek infantry soldier is small and at first he does not
impress an observer favourably.   He is, however, full of energy,
and moves lightly and easily over rough ground.   The officers
are the weak point.   Even the prospect of manœuvres had an
eliminating effect, for a number of officers preferred to resign their
commissions rather than march down from Larissa and take part
in operations in the heat.   Company commanders were mounted
for the manœuvres on small ponies.

A large number of the men, having received but little training
previous to the manœuvres, did not understand how to make the
best use of ground, and exposed themselves unnecessarily in
country which afforded ample cover.   Both in attack and defence
the men were very close together.   Attacks were made in three
lines, the formation adopted in the lines being left to the choice of
the company commanders ; sections advanced in single file, file,
or fours before forming line.

The scouting was bad, and in a country where surprise would
have been easy, adequate precautions were not taken to guard
against it.   This was particularly noticeable on the first day, when
the 1st Division had to advance along a valley and force the
passage of a saddle.   The advance was made without any attempt
to secure the heights which ran close and parallel to it.

The valise of the soldier looked particularly good and light.
Nearly the whole of the weight rests on a web band stretched
across a curved iron framework below and projecting on each side
beyond the wooden cartridge box attached to the back of the
waist-belt.   This web band presses closely against the small of the
man's back and bears the weight of the valise which rests upon
the cartridge box.   The braces are hooked to the belt and to the
valise, and the latter is easily taken on and off.

A man carries 58 lbs. when on service.   He has 100 rounds of
ammunition distributed in the two front cartridge pouches, 75 in
the cartridge box in rear and 100 in his valise.   Total—275
rounds.

Rations.—Each man carried during manœuvres one day's
bread ration and three large biscuits which were not to be eaten
without orders.

Machine guns.

The new machine guns (which are believed to be of the
Austrian pattern) have not yet arrived, and the only ones on
the manœuvres were two of the old-pattern Maxims.   The French

officers do not like these guns because of the amount of water required in a country where the supply is not plentiful, and also because of the amount of steam that escapes and thus betrays their position.

There is no doubt, however, that the French mission view English and German products unfavourably, and, unlike the British naval mission, take care that military *matériel*, which cannot be made locally, shall be purchased from their own country.

Cavalry.

The cavalry had no rôle during the manoeuvres, and the country was unsuitable for mounted action, being much broken and covered with large stones. Also, the opposing sides were already in contact when manoeuvres began. The horses, which were mostly Hungarian of a good stamp, were not in good condition, and were, almost without exception, badly broken. Horse management appears, to be as little studied in the army as it is in civil life in Greece.

The daily forage ration consisted of only 9 lbs. of barley and 6½ lbs. of hay.

The sword is carried on the saddle and the carbine slung on the back. There was a great shortage of watering places for horses owing to the water usually being found only in wells. The cavalry therefore carried wooden troughs about 6 feet long, with the regimental transport.

Artillery.

The artillery have adopted the French drill almost in its entirety, with the exception of a few modifications to suit the differences in equipment and in the capabilities of the horses. In Rumania and Bulgaria, Hungarian horses are considered to be too light for artillery work, and Russian horses are purchased instead. In Greece, however, only Hungarian horses are used, and they wear themselves out in doing with their muscles what they should be doing with their weight. This was perhaps one of the reasons why the artillery hardly ever changed position during the manoeuvres. Another reason was that they were not in touch with their infantry owing to the lack of means of inter-communication.

Batteries were usually in action in groups (each of two batteries). To each battery in a group a definite zone was allotted, and in that zone the battery commander could engage any target that he wished. Manoeuvres are not favourable for observing fire discipline, but the drill appeared to be good. There were neither signallers nor telephones, though both are to be introduced at a later date.

The mountain artillery looked workmanlike. The pack-saddles have woollen panels fitted on to hinged bars which adjust the weight evenly to the mule's back.

The component parts of the gun and carriage are carried on five mules as follows:—1, muzzle ; 2, wheels and trail ; 3, shields ; 4, carriage ; 5, breech.

The mules are of a good stamp and are purchased either
locally, in Italy, or in Cyprus. Each ammunition mule carries
12 shell—six in each ammunition box. The occupation of a
position was a leisurely movement without any of the smartness
that characterizes a British mountain battery.

Engineers.

The two engineer half-battalions were chiefly employed in
constructing a road 4 kilomètres in length. The work appeared
to have been well carried out. A number of entrenching tools
were carried on mules, but no entrenchments were made, possibly
owing to the stony nature of the ground.

Signal Service.

The want of combination between the arms, which was
especially noticeable in the case of the infantry and artillery, was
probably due to the absence of signalling equipment and telephones, although heliographs and flags are, it is said, to be
introduced in the near future. The sun shone brightly all day
and there is no doubt that the heliograph would be the most
suitable means of inter-communication in Greece.

When men only serve for 20 months, however, the supply of
trained signallers presents difficulties.

Air Service.

There were two Farman biplanes, but only one aviator. He
made ascents on two days, and then only in the early morning,
as, owing to the hot sun and the hilly country, there were many
air currents, and flying was both difficult and dangerous.

Supply and Transport.

The supply arrangements for the manœuvres were improvised
by the chief intendant of the French mission. He called together
the supply officers with the divisions engaged and drew up
provisional regulations.

In theory the troops were living on the country, and whilst
marching down from Larissa the supply officers of the 1st
Division preceded the column and collected supplies. The country
did not, however, provide sufficient to feed the small force
engaged, and the corn merchants in Thessaly formed rings and
forced up prices, which resulted in flour being purchased in Athens
and sent up to the troops by train.

Officers were permitted by law to requisition supplies if their
price was raised above current rates. Commanding officers were
given 12 centimes (1¼d.) per man, per day—beyond the ordinary
sum allowed for rations—for the purpose of improving the men's
food and for firewood. The rations included bread, meat, cheese
and coffee. Officers not serving with troops drew the same rations
as the men, but a general was allowed four rations daily, a field
officer three, and a captain or lieutenant two.

An old law was revived, which enabled transport to be re-uisitioned. The carts used were almost all of the same type, with two wheels and drawn by one horse. The wheels are high, which should facilitate draught, and this pattern of cart found favour with the French officers owing to the resemblance that it bears to carts used in Algeria. The load of a cart is 880 lbs. There were no civilian drivers with the transport. The owners of the horses and vehicles begged to be allowed to accompany them. They were not allowed to do so, in spite of the obvious advantages that would be derived from their presence, as it was held that in war they would not be present owing to the liability of un-disciplined men to panic.

One large motor lorry, made by Messrs. Schneider, of Havre, was in use. It could carry a load of $2\frac{1}{2}$ tons, but was too heavy for the roads and bridges of the country. It is unlikely that more of this pattern will be purchased.

Billets and Bivouacs.

Troops were in some cases billeted in villages, but usually the tents, of which each man carries a section, were used. The section of the tent is square, and by a system of cords, which adjust it to the required shape, it can be worn by the soldier as a waterproof coat.

Medical.

Service conditions were not practised in the medical services. Each division had one experimental ambulance wagon, of a pattern constructed locally, drawn by one horse, and capable of accommo-dating two men lying down in stretchers or eight men sitting. There were, in addition, for each division, four mules for carrying sick men in panniers and two for carrying them lying down.

Special precautions were taken to guard against malarial fever, which is prevalent in Greece, and throughout the manœuvre period every man had to take a dose of quinine before his evening meal as a preventive.

General Impressions.

Last year no manœuvres were held, as the Greek Army was considered to be insufficiently trained. This year they took place at the time when the men had received their maximum amount of 20 months' training, since it is customary to dismiss men on permanent furlough on the 1st June, four months before the expiration of the legal period. The men appeared to be cheerful and contented, and for the first time were well fed and cared for, while there were no signs of the indiscipline which was supposed to exist in the Greek Army.

The officers no longer mix themselves up in politics as they did previous to 1910. However decadent, politically, the country may have been, there appears to be no moral decadence amongst the mass of the people.

ITALY.

Every year the three permanent cavalry divisions, and the fourth provisional cavalry division to be formed on mobilization, undergo regimental, brigade, and divisional training, lasting from 30 to 40 days. This year inter-divisional manoeuvres between the 1st and 2nd Cavalry Divisions also took place, under the direction of the Inspector-General of Cavalry. The foreign military attachés were only present on the last two days, so that the following notes are necessarily brief.

Each cavalry division consited of two brigades (each of two regiments) and the following divisional troops:—

1 group of two horse artillery batteries of four guns each.
1 Bersaglieri cyclist battalion with machine gun section.
2 pack machine gun sections.
1 detachment of sappers and miners, consisting of 51 cyclist sappers and two motor lorries with stores and tools.
1 pontoon section, consisting of 48 cyclist sappers, four 4-horse pontoon wagons, and two wagons—also 4-horse—with bridging material sufficient to bridge a distance of 41 metres (43·7 yards).
1 telegraph section, consisting of 14 telegraphists on two motor lorries, each with 11 kilometres (7 miles) of wire.

The Director also had at his disposal a battalion of three companies of volunteer cyclists.

The average strength of a squadron was 85 rank and file, and of a cyclist company, 80 rank and file.

The manoeuvres were not consecutive operations based on a general idea; a new scheme was set every day. On the last day the two divisions (less one regiment) were exercised as a cavalry corps against a skeleton enemy of two cavalry divisions represented by a regiment. So large a force of cavalry had never worked together in Italy before.

Tactics.—The most noticeable feature of the operations were:—

(*a.*) Want of information. As far as could be seen, reconnaissance was left entirely to the cyclist battalions, who soon found and became engaged with the enemy's cyclists, and reported the fact, but made no further attempt to discover and report the whereabouts of the main body.

(*b.*) The want of inter-communication. The commander of the 2nd Division completely lost touch with his left (4th) brigade, which was reported to him as one of the enemy's. This fault was noticed by the Count of Turin (Commander of the 3rd Cavalry Division) who told one of the attachés that he himself always moved his division by short successive "bounds," punctuated by short halts, during which he assured himself of the position of every unit of his command. He expressed to the same officer his conviction that there was no such thing as ground unsuitable for cavalry. On the contrary, cavalry could be used with advantage in any country if the leader had studied the proper method of using the arm under various conditions.

Machine gun section with cyclist battalions.—Each gun detachment is mounted on five motor cycles, one of which carries the gun, one the mounting, and the remaining three, ammunition. The gun, a new-pattern light maxim weighing 18 kilos. (40 lb.) is carried horizontally on the top of the motor cycle. This method, which has just been introduced after many experiments, is said to work satisfactorily.

Horses.—The Sardinian horses, on which the light cavalry are mounted, are exceptionally good. They are about 15 hands high, compact, with plenty of bone, and as active as cats. They are very hardy and good campaigners, and are said to have done very well in Libya; in fact they seem to be ideal light cavalry horses. The dragoons and lancers are mounted on more "leggy" animals, many of them being Hungarians. The horse artillery had well-bred Irish horses, but lacking the weight and substance to which we are accustomed.

General remarks.—The officers, as is well-known, are good horsemen; the men also ride well, and seem to be at ease on their horses. The one defect in the system of equitation is that it does not seem well adapted to the mounted combat; indeed, practice in fighting with sword or lance is but seldom carried out in the Italian cavalry.

The material is excellent, and if a system of quicker promotion could be devised—the average age of the eight cavalry brigadiers was 57—there is no reason why the Italian cavalry should not rival that of any of the continental armies.

JAPAN.

British officers attended in 1912 :—

(*a.*) Army manœuvres.

(*b.*) Inter-divisional manœuvres of the 10th and 17th Divisions.*

(*c.*) Special bridging exercises of the engineer battalions of the 1st, 14th and Guard Divisions.

(*d.*) Exercises of the bearer battalion of the 4th Division.

Extracts from the reports of British officers attached to units during 1912 are included in the Remarks.

(*a.*)

Army Manœuvres.

The composition of the opposing forces was as follows :—

Northern Army (Red). Commander—General H. Oshima.

13th and 14th Divisions.
2nd Cavalry Brigade.
1 battalion heavy field artillery.
1 army telephone section.

Southern Army (Blue). Commander—General Y. Oshima.

Guard and 1st Divisions.
1st Cavalry Brigade.
1st Field Artillery Brigade.
1 battalion heavy field artillery.
1 army telegraph section.

48 battalions, 28 squadrons, 2 horse batteries, 48 field batteries, 4 heavy batteries, 16 infantry (6-gun) and 2 cavalry (8-gun) machine gun sections, 12 engineer companies, 5 telephone sections, and 1 telegraph section—total about 49,000 men— took part in the manœuvres.

The situation on the opening of hostilities was that a southern army had advanced from the south-west, and concentrated in the neighbourhood of Tokyo, after driving back a northern army in the vicinity.

The northern army had retreated to a line running east and west, some 13 miles northwards, and awaited reinforcements from the north. Each army was manœuvring in connection with an imaginary force on its eastern flank, the fortunes of which were swayed by the directing staff.

---

* A Japanese division consists of 2 infantry brigades (total, 12 battalions) ; 1 cavalry regiment (3 squadrons) : 1 field artillery regiment (6 batteries—36 guns) ; 1 engineer battalion and 1 transport battalion. A bearer battalion is formed on mobilization.

For reasons of economy the training of engineers and of the transport and medical departments is not carried out to any appreciable extent on army manœuvres in Japan. Each of these branches has annually special manœuvres of its own, as it is considered that the maximum of instruction and the minimum of expense are obtained in this way.

Reserves.—600 men per regiment (or an average of 50 men per company) of the 1st and 2nd Reserves were called up, and took part in regimental, brigade and army manœuvres. A large percentage had dropped out from sore feet by the time army manœuvres were reached, when companies averaged 140 instead of 170.*

This preliminary trouble from sore feet is inevitable with Japanese reservists on mobilization. When they leave the colours the majority resume the national footgear (sandals), and some time must elapse before they again accustom themselves to boots

(b.)

Inter-divisional Manœuvres of the 10th and 17th Divisions.

The 17th Division was reinforced by one battalion (three 6-gun batteries) of mountain artillery; otherwise the composition of the divisions was normal (*see* note, page 64).

The manœuvres were progressive—the phases being—inter-regimental, inter-brigade, and inter-divisional. The opening combats of each phase took the form of encounter actions, the opposing forces being started at about equal distances from the desired scene of action.

Considerable attention was paid to the passage of defiles.

(c.)

Special Engineer Exercises.

These exercises were held between the 18th and 29th September, at the mouth of the River Tone. The river at this point is about 1,000 yards wide, about 20 feet deep, and the current rarely exceeds 2 miles per hour.

The subject of the exercises was bridging. The same subject was chosen in 1910; the chief reasons for repetition, after so short a lapse of time, were that it was desired to use, on a large scale, the new pontoon and trestle equipment, and also to make a practical trial of some heavy bridging material.

The following troops took part:—

Engineers.— About 300 men from each of the engineer battalions of the Guards, 1st and 14th Divisions.

Other arms.—One infantry battalion, one squadron and one battery attended on the last two days for practice in crossing the river in rafts, and marching over the bridge made by the engineers.

---

* The normal peace strength of a company is 150.

F

The exercises were under the direction of the Inspector of Engineers. His staff was divided into five sections, viz., operations, reception (of spectators*), recording, stores, and miscellaneous.

For purposes of training and criticism the operations were divided into the following stages:—

(a.) *Preparatory.*—

> Surveying the river.
> Marking the centre-line of the bridge, and the positions of anchors, by day and night.
> Construction of shore-ends and fixed points in mid-stream, and anchoring of the latter.
> Laying out stores.
> Arrangements for rescue of men falling into the river, and precautions against accidents.
> Method of taking pontoons to their places in the bridge.

(b.) *Construction.*—

> Allotment of bridging parties.
> Command of bridging parties.
> Bridging (three methods)—
>
>> A.—Forming up.
>> B. - Forming bridge from rafts.
>> C.—Preparation and loading of every pontoon beforehand with the stores necessary for one bay, the pontoons then "coming into bridge" simultaneously (*see* page 78).
>
> Laying anchors.

(c.) *Work after the completion of the bridge:*—

> Forming cuts and reforming bridge.
> Indication of the point for the passage of river traffic, and location of observation posts up and down stream.
> Protection of the bridge against wind and waves.
> Correction of the roadway on account of ebb and flood tide, both as regards the centre-line of the bridge and the level of the roadway.
> Method of dismantling bridge in a gale.

The tactical element was absent this year, none but technical questions involved in bridge building being considered.

The exercises were divided into three phases. The first phase was devoted to elementary instruction in the use of the new material. During the second phase a bridge was constructed and dismantled twice during the daytime and once during the night, the times taken over construction being 7, 6 and 4 hours respectively. All three battalions were employed on each occasion, the commanders of battalions taking charge in turn.

---

* Nearly 200 officers attended as spectators, including representatives from all the engineer battalions in Japan.

The third phase lasted 2 days only ; the site was changed ; a covering party was sent across the river in rafts and actual bridging operations began at 7 p.m. "in the presence of the enemy." The bridge (800 yards long) was completed by midnight. (For remarks on the operations *see* page 77, " Bridging.")

(d.)

Exercises of the bearer battalion of the 4th Division.

These exercises were held in the 4th Divisional District from the 14th to the 17th August, both inclusive.

One infantry battalion and two troops of cavalry were employed in addition to the bearer battalion.

The directing staff included the principal medical officer of the 4th Division, a lieutenant-colonel of infantry, a major of the transport corps, two medical officers of field rank, and one apothecary officer. Medical officers, including some reserve officers from the different stations in the divisional district, attended for instructional purposes.

Similar exercises are held in each division during July or August of every year.

As the number of combatant troops was considered too small to exercise satisfactorily a complete bearer battalion, it is intended in future to increase the strength of the infantry to at least one regiment, and that of the other troops in the same proportion.

In Japan, bearer battalions do not exist as such in peace ; they are formed on mobilization from the personnel of garrison hospitals, and from reservists. On this occasion, the reservists had been called up about 10 days before the exercises began.

The operations were not continuous. A different scheme was set every morning and afternoon or evening. The various schemes comprised the following :—

Collection and disposal of wounded during the active defence of a position captured and held by the advanced guard pending the arrival of the main body.

Collection of the wounded of a force making a night attack on a strong position held by the enemy.

Collection of the wounded of a force attacking a position.

Collection of wounded after sunset on ground over which a stubbornly resisted advance had been made during the afternoon.

(*See* pages 85-88, *Medical,* for comments on the above.)

Direction of Manœuvres.

All tactical exercises in Japan are a series of distinct situations, the nature and sequence of which are predetermined by the director. It is considered that the instructional value to be gained from any exercise is proportional to the care and thought which have been expended upon its preparation by the director. If commanders are allowed a free hand, or, in other words, if the director does not know what is going to happen, he cannot so thoroughly make these preparations, and the lessons which he wishes to impart will be illustrated imperfectly. There are, of course, disadvantages to this system, but the Japanese consider that the waste of time and the haphazard instruction that to a certain extent, are involved when exercises develop naturally, outweigh the disadvantages of their system.

Apart from tactical training, the Japanese regard manœuvres mainly as providing a means of testing the powers of endurance of troops.

During inter-divisional manœuvres the commander of the 17th Division knew at least 2 days beforehand the course manœuvres were to take and the action that would be required of him.

An elaborate system of telephones was laid before manœuvres began, by which the directing staff regulated the advance of the opponents by time from point to point. The result was a tendency to arrive at the final timing station too early and, after waiting until the appointed hour, to race for the position.

Thus on the opening day of inter-divisional manœuvres, the 17th Division covered 16 miles between 1 a.m. and 8 a.m., when they were compelled to halt for $2\frac{1}{2}$ hours; they then covered $7\frac{1}{2}$ miles in $1\frac{3}{4}$ hours and came immediately into action. The time that elapsed from the first deployment of the advanced guards to the sounding of the "stand fast" was, in the case of brigades, 40 minutes, and of divisions never more than 1 hour.

This rapidity was to a certain extent the fault of the umpires who rarely used their powers.

Umpires.—No instances were observed of troops being informed by umpires that they were under artillery or other fire. Their chief occupation appeared to be the collection of information for the directing staff.

The captains detailed as umpires have no executive power, and can only make notes and draw the attention of senior umpires to particular incidents.

The authority of a lieutenant-colonel or major detailed as umpire only extends to the unit to which he is attached.

No special attention was given to artillery fire or its probable effect, nor was any definite value allotted to that of machine guns.

Casualties.—When a bayonet assault took place, the opposing troops charged past one another, without stopping, before contact took place as heretofore. The weaker body then marched back under the umpires' directions.

Apparently, no troops were ever withdrawn under the enemy's fire, nor did the weaker side fall back *before* an attack in superior force. Casualties were assembled at the front and marched in a body the necessary distance to the rear. This was, no doubt, in keeping with the Japanese idea that, once a man has reached a certain point, he never falls back again.

Local committees.—Members of the local committees responsible for matters in connection with manœuvres, and their employés wore red and white badges giving the particular service with which they were charged.

Claims for damage.—Claims are assessed on the spot by an intendance officer and the village headman, and usually are settled then and there in cash; this year, however, in the case of the 1st Division, vouchers were given. Claims for compensation sent in after the conclusion of manœuvres are not considered. Many farmers refuse to make any claim, no matter what the damage.

Staff.

As reported last year, practically no attention was paid by the staff to means of economizing the powers of endurance of the men or of adding to their comfort. Troops were kept standing aimlessly about at the conclusion of a long day's march, or were roused from their bivouacs to fall in long before there was any necessity to do so.

In justice to the staff it must be said, however, that they never spared themselves. Members of the personal staffs of general officers commanding armies, divisions and brigades stated that their chiefs rarely got as much as 2 hours' sleep during each of the four nights of the operations. The effects of this long vigil were plainly to be seen in the appearance and demeanour of many of the senior officers at the end of manœuvres.

Staff College students.—Staff College students in their 3rd year are no longer detailed to divisional staffs, as it was found that, with the object of giving them instruction, too much of the staff work was delegated to them, and that the experience gained by the actual staff suffered in consequence. They are now posted exclusively to the directing or umpire staffs.

Orders.—Orders were as a rule dictated by chiefs of staffs without reference to notes or maps and were taken down by representatives of units. One of these was then directed to read out what he had written. By this means the remainder were able to verify their copies.

In consequence of the training they receive, both officers and men have extraordinary powers of memorising and repeating orders imparted to them verbally.

Distribution of orders.—Orders issued by the commanders of armies were distributed only to the chief umpire, the divisional commanders, and the commanders of army troops; those by divisional commanders to brigade commanders and divisional troops; those by brigade commanders to units of the brigade.

The officer commanding the artillery of an army (*Hōheibuchō*) issues his own orders based on those received from the army commander.

The officer commanding the divisional artillery issued no written orders. Indeed, the absence of written orders in the junior commands was noticeable throughout manœuvres, officers preferring, where possible, to call their juniors round them and issue orders to each personally; this is always done in a low voice and a little apart, possibly to avoid confusion if too many orders are overheard.

Tactics of the three Arms combined.*

The tactics employed were singularly straightforward and simple. The first desire of every leader was to go straight for his opponent by the shortest road and close with him, and the possibility of the opponent undertaking any but a similar course of action was hardly entertained.

The approach.—When an encounter action was anticipated, the usual procedure was to march by night or early morning, with a view to making dispositions on the expected scene of action prior to the arrival of the enemy's main body. The marches were not, however, timed so as to allow of the position being reached by dawn, so that any enemy equipped with an air service would have had 2 or 3 hours of daylight in which to receive reports and arrange countermoves.

Infantry.

Attack.—Attacks were purely frontal and carried out in a line some three or four men deep at the moment of contact. The defenders remained in their trenches till not more than 100 yards separated them from the attack, when they made a counter-charge. Covering fire, except by machine guns, apparently was not used.

There was, as usual, a complete disregard of the effect of modern rifle fire, and attacks developed with extraordinary quickness, long before sufficient artillery preparation had been carried out. Every man's idea was to close with the enemy in the shortest possible time.

The impulse for the assault was invariably given from the rear by throwing in the reserve.

Entrenchments.—A fire trench was dug and occupied on the first day of manœuvres in accordance with the Infantry Field Fortification Manual. Some 200 yards of trench were made by a company in 2 hours without assistance, the soil being exceptionally easy to work.

---

* The following summary of portion of a report by a foreign officer recently attached to the Japanese army is of interest:—

Fronts.—Japanese divisions are capable of occupying extents of front in battle to which European commanders would not venture to assign formations of less than an army corps; Japanese officers and men are trained to assume that they can safely exceed by a considerable margin the extensions which are regarded as normal in Europe. Thus, a front of 1,550 yards of entrenched position normally is occupied by only four companies (800 men), four machine guns and six guns employed in dispersed sections; or a front of 3,850 to 4,400 yards on the flank of an army by only one division, which would, nevertheless, offer a very serious resistance. Japanese troops derive great help in carrying out this doctrine of wide extensions from their *moral* and spirit of self-sacrifice, from their skill in making use of the ground and overcoming natural obstacles, from their excellent training in close reconnaissance, and from their rapidity of movement, which greatly increases the value of the reserves; infantry battalions, regiments and brigades can cover distances of as much as 4,400 yards at a run. Moreover, the Japanese have a capacity for passing with lightning rapidity, undeterred by any consideration of the labour which has been expended, from an apparently passive and apathetic defensive to an offensive of the most vigorous character.

Traverses were made at irregular intervals, which were shorter in sections liable to enfilade. Obstacles were not constructed. Sketch maps on which the ranges to prominent objects were marked, were placed at intervals along the parapet.

Scouts.—The valises of scouts are carried for them in the battalion transport when possible.

Bicycle orderlies.—For the first time this year, two bicycle orderlies per regiment were employed. The machines were provided regimentally.

March formations.—Owing to the narrowness of the roads in many parts of the manœuvre area, infantry were frequently seen marching in threes.

Judging distance.—Ranges were estimated either by eye or from a map. There were no instruments for the purpose.

Marching powers.—The usual high standard was again exhibited by all ranks.

Machine guns (infantry).

Defence.—Machine guns in the defence were used in pairs. On one occasion they were kept withdrawn from the infantry line, in which platforms had been prepared for them in the trenches, until the enemy was within 300 or 400 yards. They were then brought up and maintained a continuous fire.

The allowance of ammunition for manœuvres per gun was 2,000 rounds.

The muzzle attachment for breaking up the wooden bullets worked well, and there seemed to be comparatively few jams.

Cavalry.

Strengths.—A cavalry brigade on manœuvres consisted of two regiments of four squadrons each, and one machine gun battery of eight guns. Squadrons were from 80 to 90 sabres strong.

Employment.—During army manœuvres, the greater part of the cavalry was used for purposes of protection, and when two divisions marched widely separated, for the maintenance of touch between them. On the opening day a small body was despatched as independent cavalry, but, subsequently, the distance between the opposing forces was always so small that there was not sufficient scope for their action.

Shock tactics were frequently, but not always judiciously, employed, and would rarely have been successful. In many parts the manœuvre area was favourable for shock action, but little advantage was taken of the fact.

On the final day a charge took place across the dry stony bed of a river, and through the river itself, which was some 3 feet deep. In spite of the rough nature of the ground there were only two casualties—horses falling with their riders. The far bank of the river was steep and difficult to climb, but was successfully surmounted, and the charge continued from there.

Dismounted action.—Dismounted action, especially on rear-guard duty, was frequently employed, but the degree of training exhibited by the men was not by any means up to the standard of infantry.

Infantry supports.—Infantry supports were sent out with cavalry as soon as contact had been established with the enemy.

Arms.—The new pattern carbine, with the bayonet folding back along the barrel, probably will be issued in the summer of 1913.

The sword is still carried on the person, but is shortly to be transferred to the saddle.

Telephones.—No regimental telephones were taken by cavalry on manœuvres. Inter-communication was carried out by orderlies.

Cavalry brigade headquarters had with it a telephone section of one officer and six men, the instruments and wire being carried on two pack horses. So far as was seen, however, no use was made of it.

Transport.—Cavalry were practically independent of field transport. They subsisted on what the country could produce and on the depôts established before the commencement of manœuvres.

Machine guns (cavalry).

The personnel of a battery was four officers, 64 non-commissioned officers and men, and 84 horses. The allotment to each gun was one non-commissioned officer, seven mounted men, one horse to carry the gun and tripod, and one to carry two ammunition boxes.

No difficulty was experienced by the machine gun detachments in keeping up with the cavalry. They were generally used in half-batteries, *i.e.*, four guns together, though on a few occasions the whole eight guns were massed. They were employed at ranges of 1,200—400 yards, to give cover to the dismounted attack of the cavalry, and, having come into action, usually fired continuously without changing position. The gun detachments carry swords, but are not otherwise armed. The gun is of the same pattern as the infantry one (viz., modified Hotchkiss).

Artillery.

Control.—The senior artillery officer on the field takes command of all artillery, field, mountain and heavy. In the case of a division acting independently, he is normally the officer commanding the field artillery regiment.

Tactics.—There is a strong disinclination to disperse artillery both in attack and defence. The splitting up of units is deprecated, and sections were only dispersed where the ground absolutely necessitated it. " Creeping " tactics in support of advancing infantry were never attempted, and, owing to the closeness of preliminary ranges, were seldom desirable. There is, however, a procedure known as " machine gun tactics " which consists of pushing a section or even single gun well forward to cover a road, bridge, or the mouth of a defile.

No single instance occurred of the occupation of a covered position. Even in the case of a long retreat of 12 miles, the artillery of the rearguard occupied none but exposed positions. The position most favoured by the mountain artillery is a crest-line with a few trees, or the front edge of a wood. The guns are sometimes hidden from view with broken branches or grass.

Officers' patrols.—Great importance is attached to officers' patrols. An officer's patrol from each brigade (3 batteries) invariably marches with the vanguard. Their duties are to reconnoitre for targets, positions and the approaches to positions when contact with the enemy has been made.

In the mountain artillery, though only three officers per battery took the field, two separate officers' patrols were invariably sent forward when an action was anticipated. Before the patrols go out, the general situation is explained to them, and likely positions are marked on the map; they are then held responsible for the selection of the most suitable position, its thorough reconnaissance and a general knowledge of the situation at the front. The brigade commander on arrival merely disposes his batteries on the position selected, without further question.

Owing to the short ranges at which fire was usually opened, and the speed with which the attack developed, no instance was seen of communication between the infantry firing line and the artillery.

Mountain and field artillery in co-operation.—In conformity with the usual topography of the battlefields, namely, a flat open plain bordered on each side by rough hills, the general principle governing the combined action of field and mountain artillery was as follows :—

The field artillery, massed in a central position, was held responsible for subduing the enemy's artillery fire; the mountain artillery, pushed well forward on the hills on either flank, devoted its attention to the enemy's infantry as soon as it presented itself.

On the few occasions when artillery moved forward to the closer support of infantry, it was the field artillery that did so, on account of its greater rapidity in limbering up, advancing, and coming into action; also, while the mountain artillery were so far forward at the commencement as to render any further advance quite unnecessary, the opening ranges of the field artillery were slightly longer and, being on the low ground, observation was less easy. On no single occasion did mountain artillery change its position during an engagement.

The 15-cm. (5·9-in.) howitzers, 1905 model, on one occasion came into action at a range of 3,500 metres (3,828 yards) from the hostile artillery. There was direct telephonic connection between the guns and the battery commander at the observation station, with an auxiliary wire laid in case of a breakdown. In addition to this, connecting links were established within semaphoring distance of one another. The telephone operator with the battery informed the battery commander as each gun was fired, giving the number of the gun.

Methods of fire.—It was stated that bursts of rapid fire were a mistake against artillery in action. In the late war, the Japanese found that when the Russians employed bursts of rapid fire, they could see the flashes and take cover, continuing their fire when the burst was over; but with a steady rate of battery fire they never knew when the next shell was coming.

Driving.—A good system for dismounting and mounting drivers at the walk, without halting, was seen. At the order " dismount," the centre and wheel drivers dismount and the centre driver goes and walks between the leaders, holding their reins in each hand; the lead driver then dismounts and goes to the centre horses. On the command " mount," the wheel and lead drivers mount at once, and as soon as the lead driver has taken up his reins, the centre driver also mounts. It was stated that this was sometimes practised at the trot with old soldiers.

Mountain artillery. Supply of ammunition.—The supply of ammunition was not practised on manœuvres. The system, so far as it has been elaborated, is as follows, but no practical organization has as yet been carried out, and there seems to be some doubt as to the distribution of rounds in the field among the various columns :—

Sources of ammunition.	Rounds per gun.
1st line, six ponies (one per subsection)	12
2nd line, 18 ponies (*danyaku shotai*)	36
Battery ammunition wagons (*chutai danretsu*)	?
Brigade ammunition column (*daitai danretsu*)	?
Regimental ammunition column (*rentai danretsu*)	?
Artillery ammunition column (*hōhei danyaku juretsu*) ...	?

The 1st line ponies accompany the guns into action, and the boxes are lowered behind the gun.

The 2nd line ponies follow immediately in rear of their respective batteries on the march and go into action as soon as they come up to the position. Thus each gun has 48 rounds behind it at the opening of an engagement.

The battery ammunition wagons follow the brigade on the line of march and, during action, come up in rear and to the flank of the gun position, as near to it as considerations of safety allow; from 600 to 800 yards is the normal distance. The expenditure of ammunition in the firing line is watched by the non-commissioned officer in charge of the 2nd line ponies (in war this may be a reserve officer), and ponies are sent up from the battery ammunition column on demand. These ponies come directly into the firing line if possible, otherwise they are halted under cover, and the ammunition is carried up by hand. One man carries one box (weight, 132 lb.) or, over good ground, two men carry two boxes between them. The ponies exchange their full boxes for empty ones and return. These empty boxes are again exchanged for full ones sent up on demand from the brigade ammunition column. The brigade ammunition column, following as 1st line transport in rear of the column, parks about a mile and a-half in rear of the gun position.

The regimental ammunition column is not yet in existence even on paper, as no regiments of mountain artillery have been formed, but when this takes place, corresponding ammunition columns will probably be provided.

On the conclusion of an engagement, all ponies with empty boxes go back to the artillery ammunition column and refill their boxes.

In addition to its 18 ammunition ponies, the 2nd line includes six spare ponies, one of which carries tools.

Horse lines.—Two systems of picketing horses were seen—

(*a.*) Posts about 2½ inches in diameter and 4 feet long were hammered into the ground so as to project about 1 foot above it, in suitable places or in lines if there was room. The head ropes of four or five horses were then tied to each of these posts, so that the horses were standing in a circle round the post.

(*b.*) The ordinary system, which is in use in the Japanese artillery, consists of ropes (stretched between picketing posts, 2½ inches in diameter and 4 feet long, and projecting about 1½ feet above the ground), along which the horses are tied by their head ropes. The head rope consists of a double rope which is carried loosely plaited together; this rope is unplaited and tied in two places along the picketing rope.

Horse management.—All the horses of every battery are weighed at the end of each month, and the weights recorded. In some batteries the average monthly weight of all the horses is plotted on to a chart, by means of which one can tell at a glance how the horses have been doing throughout the year, and what effect changes of work and diet have had on them.

Command and discipline.—The brigade commander exercises a much closer control of his batteries than is done by the lieutenant-colonel of a British artillery brigade. Minute details of routine with regard to batteries are referred to the brigade commander, and are settled by him.

The men do not perform the same duties every day, and one day a man may be a gunner, the next a driver, and the following day may march in the dismounted party.

The following is an extract from the report of a British officer :—

"The officers, including the brigade commander, were in the horse lines nearly as early as the men in the mornings. On one occasion, an officer was not in the horse lines by the time the men had begun to harness up. This was about 2.30 a.m., the time of the start being 4 a.m. His absence was noticed by the brigade commander, who lectured all the officers on the subject at the end of the march."

Observation ladders.—Experimental observation ladders were carried. The type that is considered the most successful is made in three sections, each a single timber, the dimensions of the lowest being 4 inches by 3½ inches. Total length extended, 23 feet; weight without shield, 70 lb.; with shield, 104 lb. The latter (weight 34 lb.) covers the body from the hips upwards, and has a horizontal slit for purposes of observation.

The three sections of the ladder are bolted together: one pattern is furnished with a windlass for raising them. At the foot is a spade to prevent slipping. A metal seat projects from the top section for the use of the observer. The ladder may be secured to a tree with ropes, in which case the shield is not used. When used in the open, it is kept upright with wire stays.

Experiments are being made with a bamboo observation ladder (said to be 33 feet high) and wagon.

Climbing spikes.—Each battery carries 20 T-shaped spikes to be hammered into trees for climbing purposes.

Field glasses.—A large pair of stereoscopic glasses by Zeiss, with vertical arms about a foot long for use behind cover, was carried by each battery. These were mounted on a heavy tripod in the open but seldom used. The six gun commanders all carry Zeiss glasses magnifying to six diameters.

Ammunition.—It was stated that 80 rounds per gun of blank was allowed to both horse and field artillery. The powder used, though different from the ordinary blank, is far from being smokeless, and instantly disclosed the position of guns. It is a composition of guncotton and nitrate of salt. There is no rule regarding the carriage of service ammunition on manœuvres.

Inter-communication.—Horse and field artillery batteries had no telephones, but there was a telephone section under an officer with the headquarters of each brigade and regiment. This section was provided with three pairs of instruments and nine reels of wire, each reel holding 500 metres (547 yards).

No visual signalling was seen on manœuvres, but a certain number of men in each field battery are trained in semaphore signalling.

Engineers.

At army manœuvres, engineers were seldom employed for technical purposes. On a few occasions they were employed in clearing brushwood from the field of fire, but otherwise they were simply used with the reserve as infantry.

Occupations of conscripts.—Engineers are drawn from the artisan classes of the population in the following proportions:—

$\frac{1}{20}$th of the annual contingent are blacksmiths.
$\frac{1}{6}$th   ,,   ,,   ,,   carpenters (including ships' carpenters).
$\frac{1}{6}$th   ,,   ,,   ,,   boatmen.
$\frac{1}{20}$th   ,,   ,,   ,,   masons and miners.

The remainder are artisans of miscellaneous occupations.

Bridging.

Among the criticisms made by the director at the conferences held during the special bridging exercises (*see* pages 65-67), were the following :—

(*a.*) In bridging a very wide river, it is important to arrange temporary mooring-places for the pontoons* before they are required to take their places in the bridge ; this was not done, with the result that there was a confused mass of boats clustered round the bridge, rendering it difficult to keep the centre line straight, and also hindering the free movement of the anchor-boats. In bad weather considerable injury to the bridge might have resulted.

(*b.*) Before the exercises began, the attention of commanding officers was specially called to the importance of making arrangements for rescue in the event of accidents; the preparations made were quite inadequate.

(*c.*) Battalion commanders occasionally showed a tendency to stand and watch some small piece of work instead of exercising a general supervision over everything that was going on.

(*d.*) The lights used to indicate positions in mid-stream during night bridging were generally placed too low, and were blotted out by working parties, boats, &c.

(*e.*) Two points require special practice in order to ensure night bridging being well done, viz., the marshalling of the pontoons before they came "into bridge," and the systematic laying of anchors, to avoid their cables crossing. Instructions on these two points will be embodied in the forthcoming bridging manual.

Other points that were noticed during the progress of the exercises were—

Crossing the river on rafts. Infantry.—The men marched straight on to the raft in column of fours, the distances being closed up so that the following numbers were on board, viz. :—

On a raft made from two boats of three sections each, 32 men.

,, ,, ,, five sections each, 64 men.

,, ,, ,, seven sections each, 96 men.

Cavalry.—The horses were led slowly one by one on to the raft, and drawn up in line, the men standing at their horses' heads. Straw was strewn on the floor of the rafts, and each trooper carried a handful of grass to give to his horse during the crossing.

---

* The new steel pontoons are of two sizes, viz., (*a*) an end (bow or stern) section, and (*b*) a centre section. The dimensions of (*a*) are, length, 2·7 metres (8 feet 10 inches) ; depth, ·74 metres (2 feet 5 inches); breadth, 1·3 metres (4 feet 3 inches), and weight, 168 kilos. (369 lb.). The dimensions of the centre section (*b*) are, length 2·2 metres (7 feet 2½ inches), and weight, 170 kilos. (374 lb.), the breadth and depth being, of course, the same as (*a*). The buoyancy of either (*a*) or (*b*) is 8 tons.

Two end sections and one centre section form a pontoon capable of carrying all but the heaviest guns. Two centre sections with end sections forming a complete pontoon of four sections, are used for the passage of 15 cm. (5·9-inch) howitzers and 10·5 cm. (3·9-inch) guns.

On a raft made from boats of three sections each, five men and five horses were carried.

On a raft made from boats of five sections each, 10 men and 10 horses were carried.

On a raft made from boats of seven sections each, 15 men and 15 horses were carried.

Artillery.—The draught horses were first unhooked from the limber, which was then wheeled on to the raft, shafts leading; the gun carriage, trail first, was then wheeled on to the raft, and placed in prolongation of the limber, the latter having been placed at the edge of the raft (in the case of a raft made from 7-section pontoons, the limbers of two guns were wheeled on first, and placed at the two edges of the raft). After scotching the wheels of limbers and gun carriages, the horses were embarked in the same way as for cavalry on another raft, with the exception of two horses per gun. The gunners stood on the outer side of the guns, and the drivers at the horses' heads. Ammunition and store wagons were dealt with in the same way as gun carriages and limbers.

Rafts were either rowed across the river or towed by one of the improvised launches.

Method of surveying the river.—Two or three men under an officer or non-commissioned officer actually measured the breadth of the river with wire or twine, wound on a drum; another party, working quite independently, calculated the breadth by means of a measured base line on the near bank, and angles to a point on the further bank. The mean of the two results obtained was taken as the correct figure.

Method of guarding against accidents.—Long spars were anchored, in prolongation of the direction of the current, at intervals of 150 yards, some 300 yards below the bridge. To each spar were attached at several points on either side short lengths of rope for men to catch hold of as they drifted past. In rear (*i.e.*, down stream) of each spar was anchored a boat (either hired locally or constructed from pontoons), for the purpose of picking up men as they were carried down by the stream and taking them ashore.

Method of protecting the bridge against wind and waves.—This was a problem which still baffles satisfactory solution. One suggestion was to have removable canvas covers, or lids, for the pontoons, which could be easily fitted, as the chief danger in rough weather was the swamping of the pontoons.

Method of bridging.—The third method, mentioned on page 66, was tried for the first time for comparatively short lengths of bridge, but it is, of course, very expensive in men, and can only be used when large bridging parties are available.

Method of dismantling the bridge in rough weather.—The rule is laid down that boats are never to be taken away from the bridge singly; in other words, that the bridge is always to be broken up into two-boat rafts. This minimizes the danger of upsetting.

Method of indicating the point for passage of river traffic.—Observation posts were placed 1,000 yards above the bridge and 400 yards below, thus allowing for the relative speed of boats travelling with, or against, the current. These observation posts consisted usually of a non-commissioned officer and two or three men, in a boat in mid-stream, whose duty it was to warn approaching boats of the proximity of the bridge. The observation boats kept their position by means of marks on the two banks, and patrolled the imaginary line between these two marks; at night, the marks were replaced by lamps.

Method of distinguishing stores of different units.—The ends of baulks, chesses, &c., and the gunwales of boats were painted a distinguishing colour, rendering it much easier to pick out the stores belonging to any particular unit than if the unit's number only was painted on the various stores. Thus the 14th Engineer Battalion had green as its distinguishing colour.

Special precautions adopted with non-swimmers.—Air-bladders were filled and strapped across the shoulders of these men. The bladders are not a definite article of store, and varied in size, &c., according to the unit. They have a great advantage over cork life-belts (which were also tried), in that they do not impede a man's actions in any way, and that their weight is negligible.

Clothing, &c.—All ranks walked into the river without removing boots; everyone worked in water waist-deep, as a matter of course, and subsequently continued working on dry land, or on the bridge, without paying any attention to wet clothes and water-logged boots. Occasionally, parties detailed for continuous duty on the bridge itself were ordered to wear *tabi* (the Japanese socks).

Proposed heavy pontoon.—The Technical Investigation Committee have now under trial what may be termed a semi-permanent, or heavy, pontoon equipment. It is still in a very experimental stage.

During the bridging operations a short length of light railway of 2-ft. gauge was laid upon the four or five bays constructed with the heavy pontoons. These five bays originally formed a species of fixed point in mid-stream, and were very securely anchored; after the second phase this "point" was towed to shore, and a shore-end constructed to receive it; upon the portion of the heavy bridge that thus resulted, railway wagons loaded to their utmost capacity with stones, iron bars, &c., were wheeled, the rails mentioned above being utilized for the permanent way. The buoyancy of the bridge was such that the deflection caused by the passage of such a heavy weight was practically negligible.

The following are some very rough dimensions of the pontoon:—

Length, 8 feet.
Breadth, 7 feet.
Depth, $3\frac{1}{4}$ feet (about one-third being submerged).

The pontoon was decked in with iron plates, being thus impervious to wind and waves.

Chesses, 15 feet long and 3 inches deep ; width about the same as an ordinary chess. Baulks, width 5 inches, depth 8 inches. Ribands were clamped down and not lashed.

The buoyancy of this pontoon was stated to be 20 tons.

Signal Service.

Strength.—The strength of the three telephone and telegraph sections attached to the Southern Army was stated to be 500. Assuming the same number with the Northern Army and the directing staff respectively, the total number from the Telegraph Corps at Nakano taking part in manœuvres would be, on a low estimate, 1,500.

The probable composition of each of these detachments was 200 men for the army telephone section, and 150 for each of the divisional ones.

It was stated that the whole of the Telegraph Corps was employed, and, in view of the above, this is probably correct.

In the total of 1,500 were included 200 men of the 1st Reserve (*Yobi*), called up on the 25th October, and dismissed on the day following the conclusion of manœuvres. None of the 2nd Reserve (*Kobi*) were employed.

Visual signalling was only employed in one or two isolated instances.

Field wireless installation.—A field wireless installation of the Telefunken system · was used for the first time at this year's manœuvres.

The installation was carried in two 6-horse pole draught wagons with limbers, the drivers of which were artillerymen.

Exclusive of the latter, the detachment consisted of 14 officers and men, of whom six were mounted.

The weight of the dynamo was stated to be 1,600 kilogrammes (3,528 lbs.) ; the engine was gasoline, water cooled. The height of the mast, a telescopic steel one, was 37 metres (121 feet). It was said to have a radius of action of 300 kilometres (186 miles).

The time required to erect the station was between 30 minutes and 1 hour. Messages were taken by ear ; there was no recorder. The power of the current generated was two kilowatts.

Civilian telegraph operators.—10 civilian operators from the Department of Communications were detailed for duty at the temporary post and telegraph office established at the headquarters of the directing staff.

Air Service.

The Parseval airship, the Tokugawa biplane, Mark III., and the Blériot monoplane were seen on several occasions flying over the theatre of operations, with which they had, however, no connection.

It was stated that the results of the aerial reconnaissances made by the observers in the various aircraft were remarkably accurate.

The total number of aeroplanes in the possession of the Japanese army is at present (December, 1912) five.

Supply and Transport.

At manœuvres, supply columns were imaginary and the regimental trains were not on a war footing.

Rations.—When the 1st Division marched out on the opening day of army manœuvres (the 16th November), the following rations were carried :—

In the canteen—2 cooked meals (consumed the same day).
In the valise—1 uncooked rice ration (1·9 pints).
,,   Tinned meat, $\frac{1}{8}$ lb.
,,   Salt, $\frac{2}{5}$ oz.
(Consumed on the 17th November.)
In the regimental train—1 rice ration (1·9 pints).
,,   ,,   Tinned meat, $\frac{1}{8}$ lb.
,,   ,,   Salt, $\frac{2}{5}$ oz.
,,   ,,   Solidified bean soup, $\frac{7}{10}$ oz.
(Consumed on the 16th November.)

On the evening of the 15th November rations were drawn from the regimental train and cooked. The portion for the evening meal was eaten, and that for the first two meals next day put into the canteen. The ration carried on the men was not used, because it is the invariable custom, when possible, to utilize rations in the regimental train. The uncooked rice ration is not perishable, and in any case there was no certainty that the regimental train would be available on the following evening, as in fact it was not. It was not allowed to come up to the troops again till the evening of the 17th, when it was replenished at one of the field store depots.

Field store depôts.—For the use of the 1st Division, one field store depot only was established before manœuvres, near the place where the troops were to bivouac, on the evening of the third day. In it was stored only the rice rations for the 18th November. This was supplemented by the purchase of fish, &c., by commanding officers of units, at villages through which they passed.

For the independent cavalry, and other troops whose movements could not be predicted accurately, a few minor depots for forage were established at different villages, the headmen of which were put in charge of them. The system of stocking these depots by the *Keiribu* (supply and accounts department) is as follows :—

Some 10 days before the commencement of manœuvres, an intendance officer, instructed confidentially by the directing staff as to the number of men and horses requiring rations, the date and the place, goes out to consult with the headmen of the village indicated. The latter nominates a few trustworthy farmers for the supply of rice, barley and straw, at a fixed rate, and the depot is occasionally visited by the intendance officer, in order to inspect what has been brought in.

Tinned meat, salt, and hay in bales (if required) are sent out by the intendance from Tokyo.

Storehouses are hired by the headman, who does his work without remuneration.

The system works very well, though sometimes the hay in subsidiary depots is left over, owing to the cavalry having gone unexpectedly elsewhere. Many of the village headmen had served in the army, and co-operation between them and the intendance officer was thus made easy. It is, however, invariably looked upon as an honour by the local authorities to be able to assist in making the manœuvres a success, and their work is done thoroughly and *con amore*.

Preserved meat.—Preserved meat is manufactured for use on active service at the chief provision and forage depot, Tokyo, and the branch at Ujina. A small quantity of the stock of least recent manufacture at the rate of ⅓ lb. per man per diem is issued for use on manœuvres.

Beef is the only kind of meat preserved. It is put up in cylindrical tins, each containing about ⅓ lb., for use as part of the emergency ration, and in larger tins, each containing about 2 lb., for ordinary use. The larger tins are stored 20 in a box, and the smaller, 96 in a box.

Biscuits.—These are put up in packets of four biscuits each, each packet weighing about ¾ lb. Two packets take the place of the ration of rice. 120 packets or 480 biscuits are contained in a tin. Dessicated rice was abolished when biscuits were introduced in 1909, as it was found that too much water was required in its preparation.

Salt.—The salt ration is made in oblong cakes of salt obtained from sea water; each cake weighs 6·63 drams, and is a man's daily ration. It is stored in tins, each containing 30 cakes.

Forage.—Compressed hay is in the form of bales, weighing about 82½ lb. each. Each bale is so pressed when baling that when unwired it readily falls apart in 10 slabs, each a ration of 8¼ lb.

Forage was, as far as possible, procured locally. When, however, the pre-manœuvre reconnaissance proves that the supply is insufficient, depots for baled hay are established in the manœuvre area. Corn only is carried.

Cooking arrangements.—(*a.*) In billets, a central kitchen is set up in the open for each battalion of infantry or battery of artillery. The intendance officer, assisted by the senior of the cooks of the unit, is responsible for the site. Seven large cauldrons per battalion are carried in the regimental train.

(*b.*) In bivouac, each man cooks his own ration in his canteen. When possible, however, the men get their rations cooked for them by the inhabitants, as the appearance of the canteen is permanently spoilt by use.

Transport.—The duties of transport officers with the regimental train were merely to see that it maintained the position in the column allotted to it in orders. Owing to the difficulty in making the civilian drivers understand what was required, no attempt was made to park the wagons at night.

Carts were hired under regimental arrangements by authority from divisional headquarters. The cost of hire was defrayed out of the manœuvre grant, but the rate is not fixed, and varies according to the locality.

The man leading the horse carried a badge on his left arm showing the unit to which the stores, &c., belonged. These, in addition to rations, consisted for the most part of officers' kit-boxes and cooking utensils for all ranks. The only regulation transport was that for technical stores.

Mechanical transport was not employed in any form, nor indeed would it have been possible to do so in consequence of the nature of the bridges, few of which are constructed with a view to bearing heavy loads.

Bicycles.—Bicycles were largely used by intendance officers and non-commissioned officers in moving about the manœuvre area.

Quarters.

Billets.—In preparing billets, much assistance is derived from two local associations, namely, the " Young Men's Association " (*Seinendan*), and the reservists living in the village. The *Seinendan* is composed of all those youths who will, in the course of the next year or two, come up for conscript service.

Adjutants received the orders for billets direct from the staff as soon as the outpost orders had been issued. They then rode straight to the area allotted to them, accompanied, if possible, by a non-commissioned officer from each company. On arrival at the village the mayor was called, and, under his direction, the two parties above mentioned assembled. Then, while the reservists were preparing rough sketches of the village showing each house and the number of inhabitants, the young men collected straw, fodder and rice, and put up temporary horse lines where required. An account was kept of the supplies received from each individual, to simplify the work of the intendance officer who issued billeting vouchers either that night or the following morning.

All supplies were drawn locally, but in the case of marches to and from the manœuvre area, or of rest days during manœuvres, the villages were selected and the inhabitants warned two or three months previously.

Receipts, in the form of Government drafts, negotiable at the local bank, were given to the owners of houses in accordance with the class of accommodation and the supply or otherwise of food. The rate for officers for 1st class accommodation was 1 yen (2s. 0½d.) and for 2nd class, 70 sen (1s. 5d.), inclusive of food. In billets, the amount of floor space allowed per man was from 1½ to 2 mats, *i.e.*, 6 feet by 4¼ feet to 6 feet by 6 feet.

Poverty-stricken villages were avoided as much as possible, except when operations were actually in progress.

No care was taken in concealing bivouacs. There was much noise and many lights, the fires being visible for miles.

Shelter tents.—These consist of strips of canvas about 5 feet long, and 2½ feet broad. They are carried by each man together with one tent peg and one joint of a three-jointed tent pole. Each joint of this pole is about 1 foot 6 inches long and ½ inch in diameter. The strips of canvas can be laced together, and tents of almost any shape made.

Medical.

Sanitary precautions.—Before the commencement of manœuvres, the local prefectural authorities are warned confidentially by the War Department. They instruct the headmen of towns and villages who, with the help of the police, render returns of cases of infectious diseases and their localities. A red placard is affixed to the door of any house where such has occurred, or which for other reasons is considered unfit for the use of troops.

Manœuvres as training for the medical service.—As mentioned before (page 65), practically no advantage is taken of manœuvres as a medium of instruction for the medical service. The annual divisional bearer battalion exercises (*see* below) are considered sufficient to give all the training necessary.

Red Cross stations.—Red Cross stations were organized similarly to last year.

Latrines and urinals.—In bivouac it is laid down that trenches 35 feet long, 2 feet 6 inches wide, and 3 feet 3 inches deep, are to be dug per battalion. No instance of this actually being done was observed.

Drinking water.—Uncovered tubs of water were placed in front of most of the houses in the villages on the line of march. Some were labelled "for the use of troops," and were said to contain boiled water, others were for horses.

When necessary, the iron cooking cauldrons are used for boiling drinking water.

No special apparatus or materials for sterilizing water are carried.

Disposal of sick.—The majority of cases are treated in the unit. For those unable to keep up, there is a "Treatment of sick section" (*Kwanja Ryōyōhan*), which follows in rear of the line of march. The sections, two of which were attached to a division, were especially constituted for manœuvres.

On a man reporting sick, he is examined by the medical officer of the unit, who sends him, if necessary, to the main road, there to await the arrival of the section or he is sent back to meet it. A nursing orderly of the unit accompanies the patient, or the latter is sent on a stretcher, or in a hired cart or rikisha.

Ambulance wagons.—Ambulance wagons are not brought on manœuvres; their use up to the present has been entirely restricted to bearer battalion exercises.

There has not, so far, been any question of the introduction of motor ambulances.

Sore feet.—Grease is issued to men on demand by the medical personnel of units.

Reservists, who are the chief sufferers (*see* page 65), are supposed to have special attention paid to the fit of their boots. As a matter of fact, those issued to them are old, and are distributed more or less indiscriminately, without reference to the size of the men's feet.

Official statistics of casualties, 1st Division.—The following official return of sick in the 1st Division during manœuvres should be accepted with reserve. The number of footsore cases is probably considerably in excess of that given :—

Total number of reservists ..	2,500
Total number of sick .	239

Of these latter, about 100 were treated in the divisional " sections," and the remainder, regimentally.

Out of the above total of sick, the footsore cases were said to number 86, and cases of stomach and intestinal troubles 58. About two-thirds of the sick were reservists.

Bearer Battalions.

Each division has in war one bearer battalion. It is composed of two bearer companies, two transport companies (*Sharyō-chūtai*), and a headquarter (dressing station) section. The battalion is commanded by a major or lieutenant-colonel of infantry or transport corps ; and each bearer company, consisting of two bearer sections commanded by subalterns, by a captain of infantry or transport corps. The numbers are sufficient to form 40 stretcher squads of four bearers each. A bearer company consists, therefore, of 160 bearers, with a proportion of non-commissioned officers. It is organized, for command and interior economy, as a company of infantry.

The personnel of the bearer companies consists of infantry reservists, who have been trained while with the colours as assistant stretcher bearers in stretcher drill and first aid to the wounded.

Each transport company is commanded by a captain of the transport corps, and consists of a transport section and an assistant bearer section, each commanded by a subaltern of the transport corps.

The personnel of a transport company consists of transport reservists who are employed (*a*) as drivers of ambulance wagons, or (*b*) as bearers of the assistant bearer section. The latter have undergone a course of stretcher drill, and are in the proportion of two men to each stretcher.

The headquarter section of the battalion consists of nine medical officers (exclusive of the commanding officer, and the adjutant), an apothecary, an accountant officer, and a dressing station party under the command of the senior medical officer. The equipment consists of 36 panniers (eight in the headquarter section), two boxes containing operating tables, 88 field stretchers (40 per company and eight in headquarter section), four or more tents, and a proportion of ambulance* wagons. All the equipment is carried on light, one-horse, two-wheeled carts, some 14 to 16 in number. Three panniers are loaded on to each cart, which with the drivers or grooms are supplied by the transport battalion and remain with the bearer battalion during the operations.

---

* Actual number not definitely ascertained.

A bearer battalion is divisible into two identical half battalions, called No. 1 half-battalion and No. 2 half-battalion. The panniers are in two identical sets of 18 each, for the purpose of division.

All the personnel of a bearer battalion, infantry, medical or transport corps, wear the Geneva Convention *brassard*. The assistant stretcher-bearers of combatant units do not wear this *brassard*. They wear instead a white band on the right sleeve.

During work in the field, the bearer companies carry the wounded from the fighting line to the dressing station, in the admission section of which they deposit them, taking thence empty stretchers to carry back to the fighting line. The carriage of the wounded from one section of the dressing station to another is done by men of the medical service. The wounded are removed to the field hospitals by the ambulance wagons of the transport section, and bearers of the assistant bearer section. The stretchers used by the assistant bearer section differ from those carried by the men of the bearer companies, in that they are fitted with a detachable canvas hood, and with a carrying pole and detachable iron supports, which enable the bearers to carry the stretcher from the shoulders, as is done with the Indian *doolie*.

The dressing station party opens a dressing station (two, if the battalion is divided) about 1,000 yards behind the fighting line.* A dressing station opens in four sections (1), an admission section ; (2), a treatment section ; (3), an apothecary's section for the issue of materials, &c. ; and (4), a discharge section. These sections are partitioned off from one another by ropes, and are distinguished respectively by white, red, green and black placards by day, and paper lanterns of the same colour by night, excepting that for the apothecary's section, which is white with broad bars of black. Severe cases have red, and light cases, green diagnosis tallies attached to them, either by their battalion medical officer or in the dressing station.

It is said that a new tally for wounded is shortly to replace the old red and green tallies at present in use.

When orders are given for a whole battalion to open a dressing station, only one-half of the panniers, &c., are unloaded at first, the other half being left on the carts, so that if the dressing station should be ordered to close and re-open elsewhere, one-half of the equipment can be sent on at once to the new position. If, however, the number of wounded requires it, the second half can be unloaded and taken into use on the spot.

The Intendance Department is responsible for the cooking and latrine arrangements, this work lying outside the sphere of the bearer battalion.

Medical arrangements in the fighting line.

Unless a temporary dressing station is to be opened, the temporary stretcher bearers remain in the fighting line, taking part in the action. The stretchers used by these bearers are

---

* The site of the dressing station generally is decided by the officer commanding force and notified in his orders.

carried, together with the medical equipment of the regiment, on pack animals, and not unloaded, unless it is decided to open a temporary dressing station.

Collection of the wounded.

(a.) *By day.*—The collection of the wounded is carried out very systematically. Each bearer section, under its commander, has allotted to it a definite area from which to collect cases, and the bearers, after carrying a wounded man to the dressing station, return to their own section commanders for further orders. A bearer section advances, generally in close order, until a point is reached near which it is expected that wounded will be found.

Here the order is given to prepare stretchers, and the stretcher squads extend and start work. In cases where the nature of the ground requires it, a certain number of men are told off for duty as connecting files. These men are taken from stretcher squads and maintain connection between the section commander and the bearers of the section. They also serve as guides to point out the direction of the dressing station.

When the distance between the fighting line and the dressing station is great, an exchange post is established about half way between the two. Half the stretcher squads work between this and the fighting line, the remainder between the dressing station and the exchange post.

A squad which has carried a patient as far as this point, here hands him over to another squad which takes him to the dressing station; while the former squad returns to the front with an empty stretcher.

The "number one bearer" of each squad carries a bandage bag containing spare dressings and bandages.

(b.) *At night.*—Patrols, taken from the stretcher squads are sent out to right and left to search for the wounded. These patrols return to the company commander and inform him of the number of wounded unfit to walk, whom they have found. The necessary number of stretchers are then sent out, with the patrol as guide. After leaving a wounded man at the dressing station, a stretcher squad always returned to the company commander's party. When a path crossing the path of advance is encountered, a paper trial is laid to indicate the correct direction.

The medical personnel is definitely told off to the different sections of the dressing station, and as far as possible always remains the same. When the dressing station is opened by the full bearer battalion, four medical officers do duty in each of the admission, treatment, and discharge sections, the two apothecary officers working in the apothecary's section. When the battalion is divided, these numbers are halved. On the senior medical officer issuing his orders for the opening of a dressing station, the equipment is unloaded. All the panniers are spread out in order in the apothecary's section, the tents taken to the treatment section, and a supply of wooden poles and ropes, for marking out the boundaries, to each section. The panniers are opened in the apothecary's section, and materials and appliances at once issued to the treatment section.

When necessary, the entrance and exit, and the path leading through the dressing station, are improved by the use of implements carried in the equipment.

The ambulance wagons and stretchers of the transport companies are formed up as close as possible to the discharge section. On active service, a portion of the discharge section is set apart as a mortuary.

When a stretcher squad arrives with a patient, they deposit him at the entrance to the admission section, where he is taken over by the medical personnel. Full particulars of each man, of his injuries, and as to whether he is sent to the treatment section or direct to the discharge section, are entered here in a register. Registers are also kept in the treatment and discharge sections, and the returns are eventually prepared after a comparison of the three.

A temporary table usually is made by supporting a stretcher on wooden legs. The cases are examined on this. Such as require immediate treatment are sent on to the treatment section. Others, after any necessary readjustment of dressings, are sent to the discharge section. On the tallies attached to cases sent direct to the discharge section a note is made, in the admission section, stating whether the patients are able to walk to the field hospital, or require lying down or sitting up accommodation.

In the treatment section, two tents are pitched (four when the number of wounded demands it). One is used as an operating tent, the other to accommodate patients who, after treatment, are awaiting removal to the discharge section.

In the discharge section, patients are separated into groups according as to whether they are to walk or be carried sitting up or lying down.

In order to regulate the passage of cases to the rear, an officer from the discharge section frequently visits the second tent in the treatment section, with a view to finding out the number and requirements of the cases he has to dispose of. As stretchers or ambulance wagons become available they are loaded and formed up till sufficient are ready to form a party, when they move off to the rear.

The stretchers used by the assistant stretcher sections are carried, until required, in the ambulance wagons.

In the transport of a bearer battalion are carried rations sufficient for 200 patients, in addition to what is necessary for the personnel of the battalion.

General Impressions.

Of the spirit of officers and men it would be impossible to speak too highly. Calls were made upon their discipline and physical endurance which few, if any, troops would have sustained with equal philosophy and courage.

The quick, decisive mentality of commanders, the peculiar efficiency of the junior officers, and the steady determination and single-mindedness of the men, were as evident during these manoeuvres as on previous occasions.

The following two extracts from the reports of British officers illustrate the spirit of the troops :—

"On the last day of the army manœuvres all the combatant officers went to the conference, and were absent from their units from 10 a.m. until 4 p.m. I did not attend the conference and proceeded alone right through the troops to join the battery to which I was attached at its billets 8 miles off. Everything was going on under the non-commissioned officers as if all the officers were present."

"The breech of a gun was blown out at practice on the 3rd September; one gunner was killed on the spot and six non-commissioned officers and men were wounded. The wounded at once returned to their places at the gun and endeavoured to go on with the drill as if nothing had happened."

NORWAY.

The manœuvres took place in the Province of Hedemark, between Elverum and the Mjo Lake, from the 2nd to the 7th September.

The Commander-in-Chief, Lieut.-General Krohn, directed the manœuvres. The King was present throughout.

The troops employed were as follows:—

Eastern Force.—Major-General Ebbeson.

3rd Brigade—6 battalions, 3 squadrons, 3 field batteries, 1 company of cyclists, 1 machine gun section.

4th Brigade—9 battalions, 3 squadrons, 3 field batteries, 1 company of cyclists, 1 machine gun section.

Western Force.—Major-General Bull.

2nd Brigade—10 battalions, 7 squadrons, 5 field and 2 heavy batteries, 1 company of cyclists, 1 machine gun section.

Total fighting troops—25 battalions (12,500 bayonets); 13 squadrons (1,170 sabres); 13 batteries (52 guns).

In addition to the above, each brigade had its telegraph company, engineer company, sanitary company, ammunition column, bearer company and field hospital, supply column and field butchery. The Eastern Force had 1 aeroplane and the Western Force, a field searchlight section.

Total number of troops employed, about 23,000.

The voluntary shooting clubs were invited to form a company, but only 6 members appeared.

Concentration.—One of the features of the manœuvres was the concentration (planned for the manœuvres of 1911, which were countermanded).

This involved the transport of brigades from Kristiansand, Bergen and Christiania districts, a matter of considerable difficulty, and necessitating sea transport for the greater part of the 3rd (Kristiansand) Brigade (an operation that would be impracticable in war under the probable conditions of sea power). It was stated that the transport of troops was effected without a hitch, a result of the combined efforts of the civil railway committee and a railway commission from the communications section of the general staff. Although the line from Bergen is single, the ordinary traffic was not interrupted ; the success of this operation, therefore, reflects credit on the efficiency of the committee concerned and their powers of co-operation.

The general idea was a simple one. The Eastern (or invading) Force had crossed the Swedish frontier, with orders to seize the district between Elverum and Hamar. The Western Force, operating in its own country, was directed to defend the Hedemark and drive out the enemy.

The district chosen for the operations, though for Norway comparatively flat and open, would elsewhere be called close and difficult country; it was much wooded, the open spaces being fenced and cultivated.

Direction of Manœuvres.

Commanders of the two forces were given an entirely free hand, and operations continued day and night, except that when a certain stage in each day's fighting was reached, the "cease fire" was sounded. Troops then bivouacked, but outpost lines had to be established and night operations were permitted.

On the 4th day the defending force was strengthened by the transfer of a regiment from the Eastern Force. This was the only case of interference on the part of the directing staff, and a declaration was afterwards made by the Commander-in-Chief to this effect in the Press as an answer to rumours to the contrary.

Umpires.—The umpire staff consisted of a chief umpire (major-general) with a staff of six officers; a major-general and seven or eight officers with each of the three mixed brigades, and three officers with the cavalry. The umpires' chief duties were to keep detachments and commanders informed of the situation—or as much of it as they would know in war—and to point out when hostile fire was effective. If proper precautions were not then taken, a penalty was awarded, but troops were very seldom put out of action. Owing to the numerous and indecisive combats that took place in woods, the umpiring was exceedingly arduous and difficult, and appeared to be well done. A decision was published each day as to the line established between the two forces when the "cease fire" sounded, thus—

"The Eastern Force, after fighting all day, is unable to penetrate beyond the line X—Y—Z—where its outposts are established."

Maps—Ten per cent. of the troops were provided with maps gratis, while any man could get a map on payment of 20 *öre* (2¾*d.*), and apparently every man did.

The map of 1 : 50,000, specially printed for the occasion, was extremely good, in fact so clear and large that too much reliance was placed upon it. It is probable that more instruction would have been obtained if the ordnance map of the country, 1 : 100,000, had been used.

Infantry.

On the whole the bearing of the infantry was a little disappointing. There is fine, hardy material, but some regiments seemed to lack enthusiasm. Discipline seemed, in spite of the events of last year, to be well maintained. The infantry were perhaps tried somewhat severely, not so much in marching long distances, as in hardships due to bad weather, long hours under arms and occasionally without dinners till midnight or after. As every unit had a travelling kitchen, the last hardship could have been avoided.

In most cases the men handled their rifles as if they were well accustomed to them. Sight setting was accurate, and aim was apparently always taken. The rifles appeared to be much worn, due to the practice of allowing the men to keep them at their homes. This is now to cease although men will be given opportunities of purchasing their weapons. The Norwegian is a good shot at short ranges, but no encouragement seems to be given to practice over 800 yards, nor does the country in general afford scope for it.

Bayonets were fixed when within 500 yards of the enemy and often disclosed the whereabouts of the firing line. The most interesting feature of the infantry tactics was the forest work; here units drawn from forest districts had an enormous advantage over the others. A company commander fighting in a pine wood was asked whom he had against him; he replied, "I know they are from Bergen because they are afraid to move." Obviously this is a branch of infantry tactics where training counts. The usual formation of a battalion advancing through a wood was a firing line of two companies extended to about three paces interval, while the remaining two companies moved in line of sections in single file; touch with the firing line being maintained either by sight or connecting files. The extended companies began operations by sending out patrols (generally two per company) of some two or three men each, who advanced in the required direction—often by compass bearing—and with whom touch was maintained by connecting files. It was noticeable that these patrols when led by an expert woodsman had no difficulties, but others frequently went astray, sometimes got lost altogether, and sometimes had to wait for the arrival of the company.

In this wood fighting much necessarily was left to the initiative of subordinate commanders. There appeared to be a good deal too much firing in the woods; since there was much cover and little fire effect a silent and determined advance with fixed bayonets probably would have had a better moral effect than indiscriminate firing.

The new uniform (grey-green cloth without badges to denote arm or unit) was well spoken of, looked serviceable and was invisible when troops were still. Puttees or leggings would have been an improvement as the trousers frequently got wet up to the knee in the long grass.

The new equipment—green *rygsak* of the usual pattern with side pouches for ammunition and grey-green leather belts and pouches—was in use in several units.

Each man carried 2 days' reserve rations, while his current day's ration went into the cooking pots. Each reserve ration consisted of 1 lb. 1½ oz. biscuit, 1 oz. coffee, 1 oz. sugar and about 9 oz. tinned meat. The biscuits, sugar and coffee are put together in a sealed tin, the sugar and coffee being in separate little tins which fit into a hole in the biscuits. There are nine biscuits; the outside ones having no holes.

Infantry, cavalry and cyclists carried 60 rounds of blank ammunition on the man, artillerymen and engineers—20.

Machine guns.

The Hotchkiss machine guns were provided for the first time with an arrangement for firing continuous blank cartridge. New pattern metal belts were used.

A short Hotchkiss is on trial.

Machine guns were not brigaded, but used singly or in pairs.

Cyclists.

The cyclist companies (ski companies in winter) which turned out some 130 men strong, did excellent service until the third day when the two forces came into close contact. The rifle is carried either on the cycle or on the man, and the kit on a tray at the back of the machine. It is probable that cyclists will be found of more use to the Norwegian Army than cavalry. Their action was confined to the vicinity of roads.

Cavalry

Beyond the work of reconnaissance the cavalry had little scope; all they did could have been accomplished by cyclists. They were practically confined to the roads, and as there are few open spaces in Norway, it seems that the authorities are right in limiting the cavalry to a total of 16 squadrons. It would probably pay the Norwegians to develop their cavalry exclusively on mounted infantry lines as there is little opportunity for shock tactics in Norway. There is a school in favour of this procedure.

Mounted infantry tactics alone were employed. Lack of training and the large numbers of *utskrevne* (boarded-out) horses in the ranks made the leading of three horses by one man an impossibility: firing lines were therefore meagre.

The men are fair riders, and the horses are well cared for. The latter are either Norwegian, Swedish or Danish, and useful hardy animals, though not very striking to the eye.

The men are still armed with the old pattern carbine, though the new carbine and bayonet are to be issued. The sword is carried on the off side. The equipment is a heavy one, the average weight on a horse being 18 stone.

The cavalry of each side had a machine gun section of four guns which were generally used in pairs. Guns were mounted on light one-horse carriages. One section of two Rekylgevaer (Danish pattern) guns on pack horses was on trial.

Artillery.

The nature of the country was shown by the fact that no guns fired a shot until 11 a.m. on the third day, although the forces were in contact from the first day.

Great difficulties were experienced in observation, and it is

doubtful whether there would have been any really effective fire.
Connection between batteries and infantry was in some cases
made by the artillery telephones (1 km. cable and 10 km. light
wire, capable of forming three stations). This soon gave out,
however, owing to difficulties of the ground.

A typical artillery position, seen on the third day, was as
follows:—Two batteries in action near a farm. Battery commanders, directing fire from roofs of buildings, in telephonic
communication with their batteries and with each other. Two
observation stations, one 1,000 yards to the left front connected
with the battery by flag signals (large red and white flag using
Morse code), and the other 900 yards to the front in a tree,
connected by telephone.

The panoramic sight and collimator are not yet in universal
use. A Zeiss range-finder (horizontal bar) has been issued to each
battery, and gives accurate results (error, 25 yards in 4,000 yards).

The artillery horses are hardy and active, and now nearly all
Norwegian. The artillery will in future be horsed entirely from
the country.

The two heavy batteries were armed with the 10·5 cm.
(4·12-inch) Creusot, purchased in 1909, and highly spoken of. The
gun may be carried in two positions, travelling and firing, and a
light fore-carriage takes the place of a limber. These batteries
appeared able to go wherever there was a road, and to come
into action on fairly heavy ground. They were seen in action using
direct fire at 5,300 yards, and indirect fire at the enemy's bivouac
at 6,000 yards, with a forward observation station. The gun is
sighted up to 10,900 yards. It fires common and shrapnel shell
which are carried in four-horsed wagons. Ammunition supply
from ammunition columns was not practised.

Engineers.

The engineers were much in evidence throughout the
manœuvres; they appeared to be the best troops in the army.

The bridge over the Glommen was built in 1 hour 20 minutes
by 2 companies of engineers. It was 237 yards long. (A company can bridge 600 yards.) It was a fairly simple matter as
the river was not very deep nor the current swift (2 miles per
hour). In 4 hours, 12 battalions with their 1st line transport,
5 batteries and details crossed the bridge without a hitch of any
kind. The pontoon used was the 1902 pattern (double section).

The engineers were hard at work all the time—bridging small
streams, laying temporary roads in morasses, preparing bridges for
demolition; the telegraph companies were particularly active, and
did excellent work.

A searchlight section of 2 searchlights (4 one-horse wagons)
improvised for the occasion by the engineers, was attached to the
Western Force. It was said to have done good work in warding
off a night attack on the third evening. The wagons were limbered
with the searchlight on the wagon body.

The new carbine with a long bayonet was on trial in certain
engineer sections.

Signal Service.

The telegraph company kept the headquarters of each force in constant touch with its main detachments and with the Directing Staff by connecting up with the national telephone system, which is laid on to nearly every farm in the district. The artillery telephones were in constant use, and the light cavalry lines were also put out the first 2 days, but appeared to get rather disorganized. The light steel wire does not appear to give very satisfactory results.

Air Service.

The Norwegian Army now has 2 Maurice Farman biplanes and a monoplane, but only one machine—a biplane—was available. Norway does not lend itself to aviation, and as there were but few spots where a landing could be attempted, the flying was necessarily of a bold character. The biplane flew during 4 days. On the 1st September, before operations began, it made several flights with passengers, taking Major-General Morgenstiren (page 57), commanding the 4th Brigade, to a height of 1,500 feet.

On the 2nd September, the first day of operations, the same machine procured excellent and early information of the lines of advance of the Western Force, and established their roads and approximate strength. On the 3rd September, however, when fighting began, and the troops left the roads for the woods, the information obtained was practically nil.

Supply and Transport.

Each brigade had its own transport company, which formed a supply column. All the horses were hired or had been boarded out, and about one-third of the wagons were hired country carts. This was the first opportunity of seeing the transport companies since the transport reorganization, and the staff work between units and supply columns left a good deal to be desired.

The requisitioning and calling in of boarded-out horses, and the concentration of vehicles, appears to have been very successful, and the headquarter staff are convinced that, in the event of a general mobilization, the necessary transport could be obtained without difficulty. No civilian drivers are employed.

The first line transport of units now includes travelling kitchens, mostly of an improvised nature. The usual type consists of a roughly-made box of wood, lined with hay and the top covered by a mattress. A box contains 3 tins, which are tightly shut and padlocked, and will keep a stew hot for as long as 16 hours.

Each box contains food for a section (one-third of a company), and the country cart holds 2 boxes; 6 carts per battalion are therefore needed.

Billets and Bivouacs.

Billeting is legal in peace as well as in war, but there is reluctance to obtrude on the privacy of individuals, and also great lack of accommodation. Generals and their staffs and sometimes colonels of regiments were lodged in farms by the courtesy of the owners, but in this country troops always bivouac.

Each man carried a section and piece of jointed pole of the
4-man tent, which, with plenty of straw, is fairly weatherproof
and quite warm. The men generally sleep in their sleeping bag,
Iceland jersey, night cap and light shoes, all of which are carried
on the man,

Each officer is allowed a small kit box of regulation size and
no more. Bell or small shelter tents were carried for officers in
the following proportion :—

Regiment or brigade staff	..	..		..	1
Battalion staff..	..	..	..	..	3
Company	..	..	..	..	1
Squadron	..	..	..	..	1
Battery..	..	..	..	..	3
Engineer company	..	..	..	..	3

Medical.

The brigade bearer companies carry one stretcher to four
men, and every man carries an oil lantern. But little use was
made of the medical units, and the practice of imaginary casualties
was not seen.

The dressing carts allotted to battalions appeared to be very
complete and practical.

General Impressions.

In criticizing the Norwegian army it must be remembered
that the idea of a national force is a novelty, and that the country
has not yet become quite accustomed to the burden of universal
service, light though it is in comparison with that exacted elsewhere.

Thus the National Militia of Norway with no war traditions
behind it cannot be compared with the modern army of a military
race like the Swedes, nor would it be able by reason of its
insufficient training and lack of discipline to stand against the
Swedish troops in the open. Still less is it capable of organized
offensive on even a moderate scale.

However, the rôle of the Norwegian Army is for the present a
strictly defensive one, and there is no reason to suppose that the
nation if attacked might not turn its topographical difficulties to
good account and hold an invader at bay for a considerable time.
There is plenty of latent patriotism in Norway, and the individual
man, if impatient of control, is bold and hardy, and in his own
difficult country would display all the characteristics of a good
guerilla fighter. This is an important asset, for even if the
capital, chief towns and communications fell into the hands of the
enemy, the Norwegian riflemen (*i.e.*, 90 per cent. of the male
population) might, like the Boers in South Africa, prove a nasty
thorn in the side of an invader, more especially if the campaign
were conducted during, or protracted into, the winter months.
Their chief difficulty would be the lack of food supplies, which
might bring such a campaign to a speedy termination if measures
were not taken to establish the necessary depots in places inacces-
sible to the enemy.

It is probable that these manœuvres were as successful as the country and general staff expected, and showed at all events that a considerable force could rapidly be collected and provided for at a given point (and the point chosen is an important one in relation to Sweden), and that the troops show a fair measure of training and a reasonable familiarity with their weapons.

The organization of the mixed brigades of all arms, differing in size according to the requirements of their areas, may also be said to have justified itself as the best form of mobile defence in this most difficult country, and as providing handy, compact and self-contained units sufficiently strong for independent operation.

Simultaneously with these manœuvres, the 5th Brigade carried out an exercise (also dealing with the invasion problem) in the Stjordal in which some 8,000 men took part, and which included the sea transport of three battalions from Vevlungsnes to Hommelvik.

RUSSIA.

The following manœuvres were witnessed by British officers :—

(*a.*) Army manœuvres in the St. Petersburg Military District.

(*b.*) Manœuvres of the 3rd Caucasian Corps.

(*a.*) These manœuvres took place between the 25th and 27th August. The troops that took part consisted of the Guard Corps, three infantry divisions, two cavalry divisions, and a number of flagged units. The total numbers, including flagged units, were as follows :—

> 112 battalions, 54 squadrons, 370 guns, 3 engineer companies, 2 cable sections, 10 searchlights, 1 wireless section, 8 aeroplanes, 1 dirigible, and 2 balloons.

These troops were divided into two armies, a Western (Red), and an Eastern (Blue) army, the latter being slightly inferior in size.

Reservists.—As a rule, Russian reservists go through their repetition courses in improvised units, but this year they were drafted into regular companies. The number of reservists in each company varied from 60 to 80, and the strength of companies from about 180 to 190. As far as could be seen at manœuvres, the efficiency of units was not affected by the presence of reservists in the ranks.

(*b.*) The manœuvres of the 3rd Caucasian Corps took place between the 3rd and 20th September. The troops present were as follows :—

> 21st Infantry Division : 10 battalions, 6 batteries (24 guns).
> 52nd   „   „   5   „   3   „   (12 „).
> 3rd Caucasian Cavalry Division : 3 cavalry regiments (5 squadrons), 1 battery (4 guns).
> 3rd Caucasian Howitzer " Division " : 2 batteries (10 guns).

Total—15 battalions, 5 squadrons, 50 guns, and 1 engineer company.

Many units of the above divisions were absent in Persia and elsewhere; had the whole of the army corps been present, the total would have consisted of 32 battalions, 20 squadrons, 66 guns, and 1 engineer battalion. Further, units were much below their average peace strength; the total number of men present was not more than 5,000.

The troops invariably bivouacked, the men in the shelter tents which they carry as part of their kit, the officers in light field tents; billeting did not occur.

The operations consisted of a series of disconnected exercises lasting two to three days each. The composition of the opposing sides was changed for each exercise.

Direction of Manœuvres.

Umpiring.—At the manœuvres of the 3rd Caucasian Corps very detailed regulations for umpires were contained in a manual of 100 pages, drawn up and issued by the corps staff.

The system of calculating losses was complicated, as may be seen from the following extracts from the manual:—

" Losses from artillery are estimated according to the following table :—

"An infantry company of peace strength loses the following percentage in 3 to 4 minutes from the fire of a 4-gun battery :—

Target.	Percentage of casualties.				
	At 1,100 yards.	At 2,200 yards.	At 3,300 yards.	At 4,400 yards.	At 5,500 yards.
Men standing or marching	35	27	22	10	7
Men firing kneeling	25	20	15	7	5
Men lying facing the enemy...	17	12	10	7	2
Men advancing by rushes	25	20	15	7	4
Men firing from behind cover	2	2	2	...	...
Men in emplacements behind cover ...	...	...	...	...	...

" For fire at cavalry and artillery the percentages are—

Target.	At 550 yards.	At 1,100 yards.	At 2,200 yards.	At 3,300 yards.	At 4,400 yards.	At 5,500 yards.
Cavalry—						
Standing or moving at a walk ...	80	50	35	30	25	15
Trotting	50	25	17	15	14	9
Galloping	25	12	9	8	7	4
Artillery—						
On the move	80	50	35	30	25	18
In action	...	14	10	8	6	4

" The effect of infantry fire is calculated by a system of coefficients* which represent the effect of the fire of one company on another.

"The fire of four machine guns is equal to the fire of a company.

" If the rate of fire be slow the coefficient is halved.

" The coefficient of volley firing is half that of rapid fire.

" The coefficient of a company in the open under artillery fire is one, if under cover, two."

" At frequent intervals umpires of opposing sides will interchange sketches, showing the direction and rate of fire of each company engaged, calculate the coefficients accordingly, and add up the coefficients of each side. If the sum of the coefficients of

---

* Omitted.

one side exceeds that of the other by four in the case of two battalions (or less), or eight in the case of a larger force, that side is adjudged to have a decisive fire superiority, and to cause the opposing side 33 per cent. loss in 10 minutes, a further 33 per cent. in another 10 minutes, and in ½ hour to put it out of action altogether. If the excess of coefficients is less than four or eight, the side with an excess is adjudged to cause 33 per cent. loss in 20 minutes, and to put the opponents out of action in 1 hour. If the coefficients are equal nothing happens."

The regulations give the most elaborate instructions for casualties, of which the following two paragraphs form the introduction :—

"The experience of recent wars shows that of every 100 casualties, 20 are killed and 80 wounded (20 seriously). Of every 100 wounds, 13 are in the head, 3 in the stomach, 13 in chest and back, the remainder in the limbs. 74 per cent. are caused by rifle fire, 24 per cent. by shell fire, 2 per cent. by cold steel."

"All companies, squadrons and batteries are invariably to have tickets sufficient for every man, with numbers and the nature of wounds (according to the percentage given above) marked on them."

Instructions are even laid down for cases where the horses but not the men are put out of action.

It is obvious that such regulations are far too complicated. As a matter of fact umpires interfered very little, and casualties were not practised at all.

The direction and objective of artillery fire on manoeuvres was shown either by flashing a heliograph on to the objective, or by two flags, a large red flag with a yellow stripe to indicate that the battery is firing, and a smaller flag laid out in the line between the large flag and the objective to indicate direction of fire. The colour of the smaller flag indicated the arm fired on, viz., yellow for infantry ; red, artillery ; blue, cavalry ; white, closed columns.

Tactics of the three arms combined.

General reserve.—The new Russian Field Service Regulations* lay down that a comparatively large part of a force will be kept back as general reserve on two occasions—

(a.) When it is probable that fighting may develop on one or both flanks, or

(b.) When the information regarding the enemy is indefinite.

At the army manoeuvres on the 27th August, both commanders were in possession of accurate information.

The Western army, while delivering a converging attack over a front of 13 miles, held back as a general reserve—which was intended to penetrate the enemy's centre—8 battalions, ½ squadron and 24 guns, *i.e.*, one-eighth of its infantry and artillery.

The Eastern army fighting a delaying action on a semi-circular front and with the idea of eventually delivering a counter-attack

* Published in May, 1912.

on the enemy's right, held back 12 battalions, 1 squadron and 16 guns, *i.e.*, one-fourth of its infantry and only one-tenth of its artillery.

The Field Service Regulations imply that guns should be only detailed to the general reserve in the case of a large force. From the above it is evident that three divisions are considered to form a large force.

Artillery co-operation with infantry in the attack.—At the same manœuvres, the six batteries of the 2nd Guard Infantry Division came into action in support of the infantry attack upon the Blue right flank on the 27th August in two groups, three batteries in a village on the right rear of the attackers at about 2,000 yards, and three batteries in the open on the left at about 1,700 yards. When the attacking infantry had reached close range, one of the batteries from the left group advanced along the crest of a ridge and came into action in a direct fire position, enfilading the enemy's line at about 1,000 yards. The two remaining batteries in this group soon followed, and came into action at about 1,200 yards range.

When the attacking infantry had been some 10 minutes in the final fire position, a battery from the right group galloped along the road and unlimbered on the attacker's right, at 400 yards from the enemy's trenches.

When the first battery opened fire from the ridge, the defender's position became untenable. As far as could be seen, there was nothing to prevent the left group of batteries from reaching, early in the fight, a covered position further back on the ridge. Such a position would have forced the enemy to retire, and would have rendered the infantry advance over open ground unnecessary. The choice of an exposed route by these batteries in changing position was commented on by foreign artillery officers.

The battery that made the sensational advance on the right was exposed to enfilade rifle fire during the whole time it was in movement. When it came into action it was only covered by scanty brushwood. A Russian officer argued that such things are possible when the defender is worn out with the strain of the attack. The Field Service Regulations lay down that " certain specially selected batteries must move up to the attacking troops within close range of the enemy in order to support the infantry attack."

Infantry.

The physique and marching powers of the Russian soldier are undoubtedly remarkable. He carries a very heavy and clumsy equipment, weighing, it is said, about 72 lb.

The attack.—A Russian attack was seen on the last morning of the army manœuvres. A skeleton brigade (eight battalions with guns) in occupation of a low hill on the Blue right flank was attacked by the 2nd Guard Infantry Division (16 battalions, total, about 12,000 men), which debouched from a village and wood about 1,800 yards due west of the position. The division advanced with its left on the lower slopes of a long ridge, which ran east and west; its eastern extremity commanding the defender's position at a range of 1,000 yards. As the advance progressed, the attacker's left was thrown out and forward so as to envelop the defender's right. The ground was open except for some bushes on the left of the defender's position.

The divisional commander detailed two regiments (eight battalions) to form the firing line and supports, and the remaining two regiments as reserves.

The two leading regiments issued from the village and wood in successive lines of skirmishers, about 75 to 100 yards apart; the average interval between men, up to the final fire position, was about 3 yards. The regiments in reserve followed in exactly similar formation.

During the first 500 yards, i.e., up to about 1,300 yards from the defender's position, the advances were made by squads of 8 to 10 men, in rushes of about 30 yards; from that point to the final fire position, all advances were made by individuals at the double for distances of 10 to 30 yards. As the final fire position was approached, the firing line was gradually thickened until men were only 1 yard apart.

During the 15 minutes' preparation for the assault in the final fire position, which was about 200 yards from the defender's trenches, the skirmishing lines in rear closed up, and for, perhaps, 5 minutes before the assault, the whole attacking force lay extended in seven irregular lines, covering a total depth which certainly did not exceed 150 yards.

In the assault, all the seven lines advanced 70 yards at a walk, and then broke into a run; as usual, the defenders charged to meet them with the bayonet when they were about 30 yards distant.

In the advances throughout the attack there was no bunching, and every atom of cover was utilized. Each man before starting selected his next stopping place, ran rapidly and threw himself at once flat on the ground. On the other hand, there was no attempt to practise entrenching. Casualties were not represented, so the troops were not practised in reinforcing. Supporting lines seemed, generally, too close. It is not clear why the final fire position was selected; it offered no cover or command, and the 70 yards advance at a walk would have cost the attacker dear.

The most striking feature of the attack was the neglect of covering infantry fire. The man to man advances generally resulted in the masking of the greater part of the fire of the attack. The defender's troops were well covered, and rifle fire from the plain below could have had little effect.

Fire in the attack.—The definition of ranges has been altered as follows to suit the new Russian pointed bullet :—

—	Distant.	Middle or effective.	Close or "powerful."
	yards.	yards.	yards.
Old bullet ...	2,250—1,150	1,150—350	350—0
New bullet ...	2,650—1,400	1,400—500	500—0

As a general rule there is as little firing as possible up to 800 yards from the position. Up to 1,400 yards volleys are used at special targets. From 1,400 yards to the position the firing is generally individual, and may be either slow (*ryedki*) or rapid (*chasti*). From 1,400 to 800 yards, individual fire is, as a rule, slow, from 800 to 600 it may be either slow or rapid, and after 600 it is generally rapid. The estimated rate of volleys is 8 to 10 per minute, and it is considered that up to 12 rounds a minute can be discharged in rapid individual aimed fire.

Outposts.—Some outposts that were seen on one occasion were organized as follows:—

The sentries were in groups, two men lying down and their reliefs with a non-commissioned officer a short distance in the rear. Each group of three piquets (*zastavi*) was backed by a support (*glavnaya zastava, i.e.,* "chief piquet"). A company was detailed for each support with its three piquets. About 1 to 1½ miles in rear, the reserve was posted in a village at cross roads, and sentries were thrown out on roads approaching its position. All supports were connected by telephone with the reserve.

It was gathered in conversation with officers that the line of resistance varies with the ground; sometimes the supports reinforce the piquets, sometimes the piquets fall back on the supports, sometimes both piquets and supports retire fighting on the reserve.

The scout detachment of an infantry regiment consists in peace of some 60 men and two officers. They are specially selected, particularly for their physical qualities. It was stated last year that the practice, which caused such loss of life in Manchuria, of sending forward without support the scout detachment as a whole to reconnoitre, had led to the abandonment of separate scout detachments (*okhotnichniya komandi*), and their replacement by a system in which each company had a certain number of selected scouts. The scout detachment is still, however, used as a separate unit.

Hand grenades.—The Russian Field Service Regulations lay down that infantry will throw hand grenades in the final assault. It was stated that the men would not carry these grenades throughout the attack; but that they would be handed out to them at a convenient point during the advance, at for instance 600 yards from the enemy's position.

Ammunition supply.—The supply of ammunition was not practised at the manœuvres.

The number of rounds per rifle is stated to be as follows:—

On the man	.. ..	..	120
With the regiment—			
16 Co. S.A.A. carts at 6,000..	..	=	96,000
8 Bn. S.A.A. carts at 14,400..	..	=	115,200
Total ..	..	..	211,200

For the division—

In 1st park of artillery park brigade,
24 carts, at 14,000 per cart   ..   = 336,000
                                                            ——— 23

In 2nd and 3rd parks of artillery park
brigade, 48 carts, at 14,000 per cart   = 672,000
                                                            ——— 46
                                                                        ———
                             Total per rifle   255
                                                                        ———

The one-horse ammunition carts will accompany its companies as far forward as possible, and will be filled from the 8 two-horse battalion carts which are under an officer in a central position a short distance in rear. The battalion carts are filled from the 1st park of the artillery park brigade, 2 to 3 miles in rear. The 1st park fills in turn from the 2nd and 3rd parks, which march together ½ to 1 mile behind the 1st park. Railways or transport trains of requisitioned carts will supply the artillery park brigade from the local depôts.

Equipment.—The Russians use a simple colour arrangement to assist in distinguishing units. Thus the 1st Brigade of each division wears red shoulder straps, the 2nd Brigade, blue. Each company carries a small flag (placed in the muzzle of a rifle) with a cross. The ground colour indicates the number of the regiment in its division, the upright of the cross the number of the battalion in its regiment, and the horizontal stripe the number of the company in its battalion.

The colours run—

1. Red.
2. Blue.
3. White.
4. Green.

Thus the 14th company (2nd company of the 4th battalion) of the 3rd regiment in a division would have a cross with *green* upright and *blue* horizontal stripe on a *white* ground. These flags are often extremely useful, enabling a staff officer to tell at a glance what the company is.

Machine guns.

The value of machine guns seems to be the lesson of the Manchurian war that has made most impression in Russia. In the next war each infantry regiment of four battalions and each cavalry division of 24 squadrons, will have eight guns; at present, each infantry regiment has two machine guns.

Machine guns were not seen in action at effective or long range, but the new regulations lay down that good results may be obtained by fire at infantry extended to two paces interval :—

If standing, up to 1,700 yards.
If kneeling, up to 1,250 yards.
If lying, up to 850 yards.

On the other hand, the Field Service Regulations state that, machine guns will be most used at medium and close range. They will be of special value in the defence of tactical points against local counter-attacks, and for the application of enfilade fire before the decisive bayonet assault."

All the three types of gun described last year* were seen again, but it is stated that no more of the large wheel equipment will be ordered. Apparently it has not yet been definitely decided whether to retain the pack or cart type of equipment. However, there were no new pack guns at this year's (1912) manœuvres.

A general staff officer stated that the type of gun likely to be adopted for both infantry and cavalry is the Maxim, with the Vickers, Mark 1910, tripod. This gun is described in the "Machine Gun Handbook," issued in 1912. It weighs $97\frac{1}{2}$ lb. without the shield, which weighs 16 lb. The infantry and cavalry guns are of identical pattern, but the cart which carries the gun is drawn by two horses in the infantry and by four in the cavalry.

The gun is mounted on a low carriage with three wheels, the two front wheels having a diameter of about 1 foot, the rear wheel being much smaller. The rear wheel is on a pivot and can be turned up when the gun is in the firing position, so that the trail of the carriage rests on the ground.

The gun can be fired from a lying or sitting position. In the former case the rear wheel is merely turned up on its swivel and replaced by a leather pad on which the firer rests his chest. For fire sitting, the gun is raised by two folding legs which are un-strapped from the trail, and form with the latter a tripod stand.

The carriage has an arrangement on the mounting by which the arc of traverse can be fixed. This consists of a semi-circular metal plate with holes cut in it at regular intervals. When the amount of lateral sweep desired has been determined, small stops are fixed in the holes in the plate to the right and left of the gun, which can then only traverse to the limited extent allowed by these stops.

The gun on its carriage is lifted bodily into the cart (2-wheeled) and placed pointing to the rear (so that it can be fired from the cart in an emergency). It is secured by a single clamp at the rear of the cart, and the process of taking it on and off only occupies a few seconds.

Under the driver's seat (wide enough to accommodate two or three men) is a compartment containing tools and spare parts. In the body of the cart, besides the gun itself, is carried the water-filler and one or more belts of cartridges. In two lockers at the rear end and below the body of the cart, are carried two cylinders, each of which contains four belts of 250 cartridges each.

When the gun is taken off the cart, it can either be towed behind the cart by a rope, or dragged by one of the horses taken out of the shafts, or by hand.

The saddles on the horses are adapted for pack as well as draught, so that in country too difficult for the cart, the horses can be taken out and the gun and ammunition cylinders lifted on to them, the whole operation only taking a few minutes.

---

* *See* Report on Foreign Manœuvres in 1911, pages 64, 65.

The cylinders in which the ammunition is carried are each about 11 inches long, 15 inches in diameter, and contain four tin boxes, each box holding a belt of 250 rounds. They are dragged along the ground, when coming into action, by a rope attached to either end, and roll easily over practically any surface. They are, of course, lifted over obstacles.

The shield measures 15¾ inches high by 21 inches wide. It was not on the guns at manoeuvres.

A large rattle, mentioned in the report on the 1909 manoeuvres —consisting of a cogwheel arrangement in a rough wooden box— was used to indicate machine gun fire, and it seemed quite effective.

It is stated that the tripods become unsteady after frequent firing, but, provided this can be remedied, the equipment seemed practical, and the guns were well handled.

The detachment on manoeuvres consisted of three officers and about 20 men. There were, however, seldom more than one or two officers with the guns.

The men carry carbines slung on the back, and the detachment was equipped with good glasses (Zeiss), but no range-finders were seen.

Machine-gun detachments are exercised together at a special camp each year.

Cavalry.

"*Cavalry Training.*"—The new edition of Cavalry Training, which was issued in February to replace that of 1896, contains some points that may be of interest.

It is laid down that cavalry attacking cavalry may start at a trot in extended order: at 400 yards from the enemy they should break into a gallop and at 100 yards into the charge, at the same time closing their ranks.

In an attack on infantry, the first echelons should be in single rank extended, while the rear echelons may be in two ranks in open order.

An attack on artillery will be carried out preferably in two extended lines about 300 yards apart, the first in single and the second in double rank.

The chapter on dismounted action, which has been rewritten to bring it into agreement with the "Infantry Training" published in 1908, lays down that cavalry will on occasions be used dismounted for attack as well as for defence, and the attack must be pushed home with the bayonet. Dismounting may be "ordinary," two men out of three, the horseholders remaining mounted, or "in force," five men out of six, the horseholders dismounting. When strong fire action is required and the danger of attack on the horses is small, Cossack cavalry may dismount the whole of a troop except one man. Horses are tied in pairs, the bridle of each animal being passed over its head, under the girth of the other horse and then drawn up tight on the cantle.

For the first time instructions for "lava" action have been added as an appendix to "Cavalry Training." Units of the strength of a regiment and less may adopt "lava" action against either infantry or cavalry. The action is used chiefly in reconnaissance or with the object of luring the enemy into a position favourable for attack, and each squadron or even troop may adopt a different formation as best suited to its particular task.

Lance.—The new lance is of hollow steel, about 11 feet long; it has a three grooved point, and is said to weigh 14 lbs. The 1912 estimates provided for the manufacture of 30,000, and it is said that the front rank of all regiments will be armed with lances by the spring of 1913.

The Russian cavalry trooper with slung rifle, sword, bayonet and lance certainly looks overweighted.

Cossacks.—The Caucasian Cossacks still maintain their old traditions. All men and officers are drawn from the same district, and usually have been friends and comrades all their lives. Until quite recently, Caucasian Cossacks lived under service conditions, villages being piqueted each night to prevent raids.

Thus there is a great spirit of camaraderie in a Cossack regiment. The men come from a comparatively wealthy class; it costs at least 500 roubles to equip a Cossack, since Government provides only rifle, food and forage.

Cossacks use a light thin snaffle only and wear no spurs, a loose end of the reins serving as a whip.

The men and their horses are hard and wiry. The former it seems almost impossible to tire. When on foot they usually run in preference to walking, and are ready to dance and sing at any hour of the day or night. The officers say that one of the most severe disciplinary measures they employ is to forbid all singing and dancing for a certain period.

The weight carried by the horse is not less than 18 stone, for the Cossack is practically independent of transport. He carries a *burka* (black felt cloak) rolled behind his saddle and all his belongings in a pair of large saddle-bags. Forage is carried on the front of the saddle. No tent is carried, and if billets are not available they sleep rolled in the *burka*, which serves the same purpose as the Highland plaid.

The Russian belief in Cossacks, and the Cossacks' belief in themselves, seem to have been little shaken by events in Manchuria. The Don Cossacks are more like regular cavalry in appearance and tactics. The Caucasian Cossack is still to all intents and purposes an irregular.

Artillery.

Batteries on manœuvres had four guns, but no ammunition wagons; no service ammunition was carried.

Tactics.—Positions were usually indirect. Battery commanders as a rule observed from a position close in front of the guns, and although telephone wires were laid out, fire was usually directed by flags, the code used being abbreviated Morse with two flags. One of these was white, while the other was red, yellow, or green, according to the number of the battery in a "division" (three batteries). The object of this is to avoid confusion when the fire of several batteries is directed from one point.

A foreign artillery officer who recently visited the ranges at
Luga and near Moscow made the following statement :—

"The Russian artillery at present fires badly, but the officers
are working hard and the arm is certain to improve. When
time was available at practice camps, the most careful reconnaissance and preparations were made before a battery came into
action. These preliminaries frequently occupied hours. The
scouts were accompanied by four men who were specially trained
to make panorama sketches. An officer after seeing that
prominent points on these sketches were lettered similarly, handed
one to the commander of each battery and one to the commander
of the "division" to facilitate the direction of fire."

Ammunition supply.—When a collision with the enemy appears
probable, the 1st park of the artillery park brigade joins the
rear of the fighting column, while the 2nd and 3rd parks follow
at the head of the train.

An artillery "brigade" (two "divisions" of three batteries each)
in action obtains ammunition from the wagon bodies which are
drawn up on the left of the guns. 500 or 600 yards in rear
and to a flank of each battery are the eight gun and eight wagon
limbers. A short distance further in rear, the eight remaining
battery wagons are grouped together in two "divisional" reserves
of 24 wagons each. 2 or 3 miles in rear is the 1st park of the
artillery park brigade, and about one day's march further back are
the 2nd and 3rd parks.

The artillery parks refil from depots which are without
service transport. When the distance becomes too great for the
parks to fill from a railway (in touch with a depot) the interval
will be bridged by requisitioned transport. For some years to
come the bulk of this requisitioned transport will be horse drawn.

The commander of the artillery park brigade remains with the
commander of the artillery "brigade" in action. The latter fixes
the place for the 1st park, and the park brigade commander the
place for the 2nd and 3rd. Echelons of ammunition supply are
linked by orderlies or telephone.

The ammunition carried for the Russian field gun is as
follows :—

With the battery—

		Per gun.
In 9 limbers (1 spare), 36 × 9 } total, 1,028 ..		128
In 8 wagons, 88 × 8 .. .. }		
In "divisional reserve," 24 wagons (88 per wagon)		88
In 1st park of artillery park brigade, 34 wagons (92 per wagon)		65
In 2nd and 3rd parks of artillery park brigade, 68 wagons of (92 per wagon)..		130
Total		411

Guns for use against aircraft.—It is stated that the Russians
consider that the ordinary howitzer will be sufficient to contend
with aircraft, and that no special weapon will be made.

Observation of fire.—Balloons were again used this year for the observation of fire. Aeroplanes are said to have been tried and found unsuitable.

Observation ladder.—The Russians are at present without an observation ladder. The old one, which was drawn in a 2-horse wagon, has been discarded as too heavy. The Artillery Department is supposed to be experimenting with the object of finding a new pattern. The delay in finding a new pattern was the text of one of the attacks on the administration of the department in the Duma last spring, when M. Guchkov recommended the adoption of the German type of ladder.

The director is of very similar pattern to our own. Batteries are well provided with field glasses of Zeiss pattern.

Every unit, infantry, cavalry, artillery, carried two or three torches. They consisted of tins, containing kerosene sufficient to burn all night, on the top of 8-ft. poles. They were placed in the horse lines of mounted units at night.

Engineers.

Personnel.—The best educated men and those with a trade are selected from each conscript contingent for service in the navy and the engineers. It is for this reason that revolutionary outbreaks are more frequent in the navy and in engineer battalions than in army units recruited from uneducated peasantry.

Pontoons.—There is no present intention of introducing a pontoon of sufficient weight-bearing capacity to carry the heaviest mechanical transport. The Russians are satisfied that their pontoon is a good one, while the introduction of mechanical transport is not a question for the immediate future.

Signal Service.

Visual signalling.—A regimental officer stated that he had 10 expert signallers in his company, and that most of his men could read Morse slowly. A general staff officer stated that the semaphore system had been discarded in the Russian army.

The only visual signalling seen was the passing of short conventional signals with flags of the size of semaphore flags, and coloured to represent regiments and battalions.

As the Russians have 13 mounted orderlies per regiment, and excellent telephones, they do not attach much importance to visual signalling except in mountain warfare and for the passing of conventional signals between the skirmishing lines in an attack.

Experimental Zeiss lamps for signalling by day as well as by night, and electric lamps for night signalling, were used.

Wireless telegraphy.—The Russian army is believed to have 110 cart sets of wireless either delivered or ordered. These sets are organized in sections and are manned by engineers.

A saddle station and a small cart station, with a radius of 40 to 45 miles, which were purchased privately by the Guard Hussars, are said to have proved a great success at manœuvres, and the possibility next year of large government orders for sets to issue to cavalry regiments is being discussed. The desideratum seems to be a light set in a small two-wheeled cart, which could be removed from the cart and carried forward on pack horses in case of necessity.

No arrangements were made at manœuvres to prevent opposing wireless installations from interfering with one another, and it seems probable that in the army manœuvres, at all events, the wireless will be used only by the directing staff.

The Russians purchased, in 1912, an automobile, fitted as a wireless station. The car is described as of 20 h.p., four seated, and with a collapsible mast. It has a power of 2 kilowats, and an actual radius of 100 miles. It was supplied by Siemens. Tenders have been invited for the supply of three more cars, at a cost of 2,400l. each, and the specification insists on each car having an extra engine, in order, apparently, that the apparatus may be worked during temporary breaks down of the car engine.

The dirigible, at manœuvres, was said to be fitted with wireless, but no details were obtainable.

Telephones.—The telephone is very widely used, and is invariably laid out during movement, *e.g.*, to connect the advanced guard to the main body, and in a long marching column the line runs down the column from unit to unit.

The method of laying it out is as follows: Two men, one of whom carries a drum of wire slung across his back by a leather strap, and the other an instrument, march together. The man with the instrument is connected to the wire and has an earth connection by means of metal shoes on his heels. He holds the receiver to his ear the whole time, and, theoretically, can carry on a conversation while marching. The drum of wire on the other man automatically unwinds as he moves forward. Behind this party march other men, one of whom is responsible for the wire being unrolled, putting it to the side of the road, preventing it from getting hung up, &c. Others carry spare drums, instruments, &c. When a drum is finished (each drum has 1 verst —1,100 yards—of wire), the man remains behind with the empty drum, and another takes his place, and is connected (the process takes about a minute). At the other end the reverse process is going on. The man with the drum carries it slung in front of him instead of over his back, and winds up as he advances; the man responsible for disentangling the wire marches some 20 yards in front with a pole with a hook on it. When a full drum has been rolled up, a man with an empty drum, dropped from in front, is met, and takes up the work of rolling up. The full drum is sent on again (by mounted orderly, if possible) to the other end of the wire. This sending on of the full drum, so as to ensure a constant supply, causes some difficulty. There are no mounted signallers, so, if no mounted orderlies are available, it is difficult to pass the wire up quick enough, and if the wire is sent on by mounted orderly, the man to carry the drum is left behind, and has to get to the head of the column as quick as he can on foot.

Cavalry carry the same telephone equipment as infantry, and run out the line in similar fashion. An earth connection to the shoe of the horse, on which the man with the receiver rides, is provided by a wire run down the horse's fore legs, in a pair of canvas trousers.

At the manœuvres of the 3rd Caucasian Corps a sketch-plan showing the signal arrangements was always issued with orders.

Air Service.

There are now believed to be 50 or 60 aeroplanes in the Flying School at Gachina. 150 Nieuports at an average price of 900l. have been ordered, 70 from the Dux Company, 25 from the First Russian Aviation Company, and 55 from the Russian Baltic Works at Riga. Delivery is to commence in September and to be completed by the spring of 1913.

Supply and Transport.

Preserved meat.—Frozen or preserved meat is not at present used in the Russian army as it is said to be disliked by the men.

According to the Field Service Regulations, in time of war cattle will be driven with the train together with "sutlers' carts and private carriages." The drove for each army corps will be from 250 to 300 head. A recent army order describes a method of preserving meat by injecting a solution of salt and saltpetere after killing.

Hay.—On service it is stated that 2 days' supply of hay will be carried in the regimental train, and half-a-day's supply on the horse or cart.

General Impressions.

The Russian soldier has magnificent physical qualities, and can stand great fatigue, his requirements as regards food are simple, and as regards comfort practically nil; he is obedient, and would probably be a great deal less subject to nerves than some of the more highly educated European armies.

There is an absence of anything in the nature of "spit and polish" among the Russian soldiery. Smartness, such as we know it, can hardly be said to exist, even under the best and most normal conditions.

The officer has many good qualities, but his lack of education and the poorness of his prospects are fatal at present to any great improvement. Few officers in the infantry or artillery have any means but their pay; promotion is slow, and life in a small permanent garrison hardly conducive to efficiency.

It is a mistake to consider that there is a large amount of drunkenness amongst officers of the Russian army. The number of *regular* hard drinkers in a regiment is small. The senior officers drink very little, and scarcely any of the younger subalterns of a regiment to which a British officer was attached drank anything at all.

Relations between senior and junior officers off duty are extremely friendly.

A great many of the younger officers are interested in their profession and keen, but the Staff College is open to few and is practically the only outlet for the ambitious. So the keenness is usually soon dulled by the routine of a conscript army.

The staff officers were keen and hard working. Their chief faults are the " casualness " and lack of business instinct inherent in the Russian character, and the tendency to prefer a perfect scheme " on paper " to a less ambitious one which is more likely to stand the test of practical use. On the other hand, the Russian has a greater talent for improvisation than he is usually given credit for.

SWEDEN.

The manœuvres were carried out this year in Västergötland on a considerably larger scale than usual; they lasted from 7 p.m. on the 1st until 10 a.m. on the 5th October. The forces engaged were as follows :—

Red Army.—Commander—Lieut.-General A. F. von Matern.

1st and 3rd Divisions.*
Hussar brigade (8 squadrons and 4 guns).
1 heavy artillery regiment (8 guns).
1 howitzer brigade (8 guns).
1 field telegraph detachment,
    and line of communication troops.

Blue Army—Commander—Lieut.-General G. F. O. Uggla.

2nd Division.*
13th (Composite) Brigade (6 battalions, 1 squadron, 3 batteries).
Dragoon brigade (8 squadrons, 4 guns).
3 heavy batteries (12 guns).
2 searchlight sections.
1 field telegraph section,
    and line of communication troops.

Total strength—approximately, 38,000 men ; 8,000 horses and 128 guns.

Guide officers.—Officers who were well acquainted with the topography of the district were attached to each army and divisional headquarters. They were termed " Guide officers."

Country.—The country in which the manœuvres were carried out is sparsely inhabited, rocky and thickly wooded. The open spaces are divided into small fields by stone walls or wooden fences. The ground in the interior of the woods is as rocky as in the open spaces, so that it was almost impossible for mounted troops to move off the roads; while, generally speaking, the view was so restricted as to make artillery action exceedingly difficult.

Nature of the operations.—The general idea was that the Blue army, representing a force detached from the main Home army which was concentrated on the Swedish-Baltic coast, had been despatched to Västergötland to oppose the Red army which had landed at Göteborg.

The intentions of the Blue commander were to attack the Red army before it could be reinforced, and in any case to protect an

---

* The normal composition of a Swedish division is 12 battalions, 4 squadrons, 12 batteries (48 guns), and 1 machine gun detachment, besides non-combatant services.

important railway junction which was required for the concentration of other Blue troops. The Red commander's plan was to defeat the Blue army and seize this junction.

The Blue army was eventually forced back and took up a position covering the junction. The Red commander followed, and directed almost the whole of his force (17 battalions) against the enemy's right flank. The Blue commander, obtaining early intelligence of this movement, left a brigade to meet it, and moved the remainder (one division) against the enemy's right and rear. The two armies were thus circling round one another when the manœuvres were broken off.

Concentration.—Very few units reached the area entirely by route march, practically all having had to rail at least part of the way. The concentration of 38,000 men and 8,000 horses caused considerable trouble on the Swedish railway system, as numerous private lines had to co-operate with the State railway. The traffic difficulty was increased by the fact that the Swedish lines are single. Special arrangements had to be made for on-and off-loading and for illumination, as most of the work had to be done at night to avoid disturbing the regular traffic. 108 trains were required.

Direction of Manœuvres.

The manœuvres were continuous throughout. Commanders on both sides were allowed entire liberty of action after the initial disposition of troops had been made.

The King, assisted by Lieut.-General Bildt, the Chief of the General Staff, acted as Director.

The members of the Defence Committee* were also present.

Umpires.—The service of umpiring appeared to be efficiently performed and, notwithstanding the wooded and enclosed nature of the country, umpires were nearly always at hand when required.

They were distinguished by ribbons of blue and yellow (the Swedish colours) on their caps.

Umpires were organized as follows:—

Chief umpire (to each side) with the following staff:—

7 officers.
6 mounted and 4 cyclist orderlies.
2 signalling squads.
1 chauffeur.

Three infantry groups (each 3 officers and 3 orderlies).
Two cavalry groups (each 5 officers and 6 orderlies).
Two artillery groups (each 1 officer and 2 orderlies).
Two transport groups (each 2 officers and 1 orderly).

* The Defence Committee was appointed by the present Radical Government about a year ago to investigate conditions of Swedish defence. It consists of four sub-committees, each composed of three radicals, one social-democrat and one conservative. There are no naval or military members.

An umpire accompanied each larger unit, such as an infantry regiment (3 battalions) or an artillery brigade, and was in every case a field officer. To each of these umpires were attached an adjutant (captain or subaltern) and several mounted orderlies, signallers, and telephonists with a portable telephone set. Umpires were thus able to watch the progress of an action and at the same time to keep in communication with the chief umpire and with umpires on the opposite side.

The following is an extract from the rules with regard to the indication of targets by artillery :—

"The target at which artillery is firing will be indicated on all occasions by means of screens painted with blue rectangles or circles on a white ground, according to the target.

"The screens will be put up in pairs, one slightly higher than the other (with a distance of 25 to 50 metres between the two, the lower screen being nearer to the target), in the direction of the line of fire, at a distance of not more than 1,000 metres in front or in rear of the artillery firing. Both screens will be placed so that they are visible to as large a portion of the target as the ground and other conditions permit.

"Firing against cavalry on their horses will be indicated in the same manner as against infantry, but the screens will be turned to one side.

"If the screens remain unmoved, without shots being discharged, it will be understood that firing is proceeding.

"The personnel detailed to look after the screens are neutral and must not report to their own side any observations made regarding the enemy."

Infantry.

There is little to add to what has already been written in previous reports. The men are cheerful, amenable to discipline, active, and extremely hardy. They are not accustomed to work in large bodies. Operations on the manœuvres almost invariably resolved themselves into isolated combats between small bodies of skirmishers, and for this form of warfare Swedish infantry appear to have a natural aptitude. They display considerable skill in advancing or retiring under cover; they take every opportunity of entrenching, but do not construct elaborate works.

An extension of more than one or two paces either in attack or defence was not witnessed. The infantry attack was not made in sufficient depth.

Fire discipline was good, and the control of fire more up to date than that of the German Army. The French " finger method" of indicating targets was employed. The men are alert and interested in the proceedings, ready at all times to take advantage of any fleeting targets.

Ammunition and tools.—In war, 120 rounds are carried on the man, *i.e.*, 90 in the belt and 30 in the haversack. Each of the 4 ammunition carts of a battalion contains 17,900 rounds, and a number of tools (4 picks, 6 jointed saws, 1 spring saw, 1 hatchet and 1 spade), and each of the 17 carts of the divisional ammunition column contains 23,760 rounds.

Each company has 3 wire cutters, 12 jointed saws, 48 hatchets and 100 spades.

Rations.—Each infantry soldier carries the unexpended portion of the day's ration and 3 reserve rations.

Transport. — The transport of infantry formations is as follows:—

> Regimental headquarters—1 intendance wagon, 2 baggage carts, 13 supply carts.
>
> Battalion headquarters—1 medical cart, 4 ammunition carts, 1 meat wagon, 1 baggage wagon.
>
> Company—1 baggage wagon.

Machine guns.

The peace establishment of the Swedish Army includes only one machine gun battery, which is manned by the garrison artillery, but it is proposed to provide each infantry regiment in the near future with a machine gun company somewhat similar in organization to the new German formations.

Cyclists.

Cyclists are only used as orderlies (except on the island of Götland, where there is a cyclist company). Cyclist orderlies carry the rifle upright in clips on the front stem of the machine.

Cavalry.

The cavalry had a very difficult rôle, as it was almost impossible for this arm to move off the roads.

As soon as two cavalry bodies came into collision along the roads the men dismounted, extended in the wood—which was always close at hand—and advanced against each other in skirmishing order.

The horsemanship of the Swedish officers is remarkably good, and they are well mounted. The good quality of the Swedish Government horses has been mentioned before.

Equipment.—The total weight carried by a cavalry horse is about 271 lbs.

One blanket is carried under the saddle, and there are large wallets on the front of the saddle. The wallet on the near side is divided into two compartments, the upper half containing the man's personal effects and the lower half, a canteen. The bottom of the wallet can be opened and the canteen taken out.

On the back of the saddle are strapped a corn sack, similar to the German " *Futtersack,*" a greatcoat and a picketing rope.

The cavalry soldier carries for himself and his horse the unexpended portion of the day's ration and one reserve ration.

Each squadron has—

3 baggage wagons.

1 ammunition pack horse (carrying 2,560 carbine cartridges and 625 revolver cartridges).

1 pack horse for tools (which include 4 spades and 2 boxes of explosives).

Cavalry pioneers.—One corporal and eight men per squadron are trained as pioneers and each carries a hatchet and wire cutters.

Artillery.

The wooded nature of the country made the work of the artillery very difficult and, in many parts of the manœuvre area, impossible. Indirect positions appeared to be the rule, and entrenchments were usually made for the guns.

A British artillery officer who was attached to a field battery writes as follows :—

"I did not see any control of larger artillery formations than a single brigade ; the brigade to which I was attached forming the artillery of the composite brigade. The lieutenant-colonel commanding was always in touch with his infantry brigadier, and the control which he exercised over his batteries resembled British procedure. In fact, Swedish artillery tactics closely resembled our own, and battery commanders, when once in action, practically were given a free hand.

Manœuvre.—" Batteries invariably came into action behind the crest, either in covered or semi-covered positions. In the latter case teams were never exposed, as the guns were un-limbered under cover and run up by hand. Batteries always came into action in column of sections, close interval, the wagons left (or right). No markers or range-takers gave away the position before it was occupied.

"The Swedish method of draught (the German system) is not so well adapted to quick movement as to rough going. Spring draught is used as in France, but the low pole did not strike rocks and inequalities in the ground as might be imagined.

Fire tactics.—" The Swedish battery has only three officers, the captain and two section commanders, and one of the latter is frequently absent on reconnaissance. As in the French Army, the firing battery consists of four guns and six wagons, the two extra wagons being placed on the flanks and to the rear of the battery alignment. The ammunition wagons are placed on the right of the guns, as the loader is on that side.

"The battery commander's staff is nearly as large as with us, but there are no range-finders or plotters. Observation ladders have been tried, but have not been adopted, as it was thought that they gave away the battery position. The battery commander's director has a forked stereoscopic telescope similar to the German one. It is provided with graticules,

as are officers' binoculars. It also has a level for measuring
the angle of sight, but this was seldom used; even the regulations suggest that the angle of sight should be estimated by
eye. The Swedish gun (6·5 cm.—2·5-inch) manufactured at
the Bofors Works) is not provided with an independent line
of sight, and the sighting gear resembles the Krupp arc-sight.

"An aiming point was used almost always, and the
procedure was simple and rapid. Swedish fire tactics seemed
to be a compromise between the French and German systems.

" Distant observing stations were employed, although they
were by no means the rule. On one occasion I saw a captain
directing the fire of his battery from a distance of 800 yards.

" Whenever time allowed, guns in exposed positions were
screened with bushes and branches of trees.

" The gunners are armed with carbines (carried slung over
the back), and did not seem to be hampered by them.

Transport.—" The transport of artillery formations is as
follows :—

> *Regiment* (4 brigades).—" 2 baggage, 1 intendance and
> 16 supply wagons
>
> *Brigade* (3 batteries).—" 1 tool wagon and 1 baggage
> wagon.
>
> *Field battery.*—" 10 ammunition wagons (6 with the battery),
> 1 tool cart, 3 baggage and 3 supply wagons.
>
> *Horse battery.*—" 4 ammunition wagons, 1 tool, 2 baggage
> and 2 supply carts.

Tools for 1 *battery.*—" 31 spades, 13 picks, 10 crowbars,
13 hammers, 17 hatchets, 3 spring saws, 3 jointed saws, and
17 billhooks.

Ammunition.—" Each officer, non-commissioned officer and
driver carries 25 revolver cartridges. Gunners, who are armed
with a carbine, carry 40 rounds per man. There are 284 rounds
per gun with a field battery, *i.e.*, 44 in each limber and 240 in
the ammunition wagons. (Each of the 10 ammunition wagons
has 48 rounds in the wagon-body and 48 in the limber.)

" In a horse battery there are 132 rounds for each gun,
i.e., 44 in the limbers and 88 in the wagons.

" Each of the 16 wagons of the 3 divisional ammunition
columns contains 96 rounds.

Inter-communication.—" I think we have a great deal to
learn from the Swedes as regards telephones. The efficiency
of telephonic communication in their batteries was, I think,
due to better instruments and more efficient operators. Each
battery carried 1,500 metres (1,640 yards) of wire. It was
uninsulated, but this did not seem to affect the working of
the telephone, even when the ground was sodden with snow
and rain.

" As telephones are universally used in Sweden, all the
men are accustomed to them, and do not require special
training. I did not hear any of the constant calling-up, queries
and repetition that too often occur with us. Even when

the battery came into action rapidly, the telephone was always ready for use before the first round was fired.

"There is no cable cart for an artillery brigade, but the brigade staff include mounted telephonists, equipped in the same way as the battery ones. The responsibility for communication is upwards, *i.e.*, the battery is responsible for laying out the line to brigade headquarters, and brigade headquarters to the higher artillery commander.

"Flag signalling (semaphore) is used a good deal, chiefly in case of failure of the telephone. Morse signalling with large flags is also practised, but not frequently. The flags used are exactly the same as ours.

Co-operation with other arms —"Responsibility for the maintenance of communication between artillery and infantry rests with the artillery brigade commander. The means suggested are officers' patrols, sent by the artillery brigade commander to the infantry commander with whom he is working. This was the method (occasionally supplemented by telephones) which I saw employed."

Signal Service.

Woods and bad roads made inter-communication extremely difficult. Swedish troops, however, are specially trained in wood fighting, and as a rule, show no little skill in maintaining cohesion during movements through forest country.

Signal detachments.—The service of inter-communication is mainly in the hands of "signal detachments" (very similar to the British signal companies), one of which is attached to each regiment of infantry and cavalry. Each detachment includes orderlies, signallers, telegraphists, and telephonists, who are furnished with the necessary equipment, and also with special appliances for tapping telegraph and telephone wires. These appliances were much used, as with certain restrictions, the existing telegraph and telephone systems were available for army use.

Wireless telegraphy.—A wireless equipment (Telefunken) was on trial for the first time. The equipment for a station was carried on four pack-horses. There were, it is believed, four stations, two for each army.

Each station had a detachment consisting of one officer, one non-commissioned officer and six men, and a radius of 300 kilometres (187 miles).

A German civilian from the firm which furnished this equipment accompanied each station, and was the only man who knew how to work the apparatus. It was, therefore, of little use. The equipment appeared to be clumsy.

Notwithstanding these modern methods and appliances, the service of inter-communication did not always work satisfactorily. The commander of the Red army was heard to say that, though he had at his disposal mounted and cyclist orderlies, aeroplane, signallers, telephones, telegraphs, &c., he was out of touch with his southern detachment throughout the first day of the manoeuvres.

Air Service.

There are only two aeroplanes in the Swedish Army, an 86-h.p. Breguet biplane and a 50-h.p. Nieuport monoplane; both were allotted to the Blue force. (There is also a dirigible, but this did not take part in the manœuvres.) Baron Cederström, a civilian aviator, volunteered for service during the manœuvres and was attached to the Red force. He flew a Bleriot.

Aviation on the manœuvres may be said to have been a failure. The weather and the ground were no doubt most unfavourable. The wooded nature of the country made observation extremely difficult, and landing places were few and bad.

Both the aeroplanes allotted to the Blue army were damaged and rendered useless before the end of the manœuvres without, it is understood, any satisfactory results having been achieved. The Red commander complained that his aviator, Baron Cederström, being a civilian, was unable to appreciate the military value of his observations ; he was not accompanied by a military observer. He did, however, furnish one report of value, as he was able to state on his return from a flight that a certain named area was free of any enemy.

Od the 3rd October the same aviator was in the air for 45 minutes, but as the landscape was covered with snow he could neither find his own bearings nor locate the troops observed.

Supply and Transport.

The troops always were supplied with fresh bread and meat. Each battalion had a wagon like a butcher's van, from the roof of which the meat was hung. These meat wagons belonged to the 1st line transport, and were provided with poles and guys for hanging freshly-slaughtered animals. Preserved meat was carried for issue in case of emergency, and to add variety to the diet.

No travelling-kitchens were seen, but it is believed that they are to be introduced.

It was not possible to obtain any supplies locally for the troops. Supply depots and bakeries were, therefore, established in two places, and troops were fed by means of supply columns. A number of motor wagons were hired to assist in the service of supply, as there is no mechanical transport permanently in the hands of the military authorities. All army baggage wagons were of the limbered type. (*See* also under MEDICAL.)

Horses.—The horses, both riding and draught, were of excellent quality. Although only about half the number that took part in the manœuvres belonged to the State—the remainder being registered and called up for the manœuvre period—all seemed to do their work and stand the strain equally well. The severe cold, wind and snow to which they were subjected appeared to do them no harm, but the excellent Government horse-blankets entirely cover up the little Swedish horses from ears to tail. They always stood very quietly, and heel-pegs were not used.

The reason why the horses stand the work so well is probably due to the fact that they are given no hard work until they are

six years old. Remounts are purchased between the ages of three
and six; spend one year at a remount depot; and then undergo a
further period of two years' training on joining their unit, before
being put to regular work. The price has hitherto varied between
£33 and £55, but this has just been raised to £46—£52 for three-
year-olds, and a maximum of £61 for four-year-olds.

Billets and Bivouacs.

The troops for the most part found shelter. at night in sheds
and barns, but on several occasions the men slept in the shelter
tents they carry with them.

There is no billeting law in Sweden, so the troops are dependent
on the goodwill of the population. The price of a lodging is not
fixed, but usually is 10 *öre* (1¼*d.*) for a man, and 25 *öre* (3*d.*) for a
horse per night. The Government refunds commanding officers.

Medical.

As in the manœuvres of 1910, field dressing stations were
established at various points, and admirable preparations were
made for the reception of the wounded. The treatment of
wounded was practised by attaching labels to men stating the
nature of the supposed injury. The men were then removed to
the dressing stations and treated accordingly.

The ambulance wagons are of a very light pattern and consequently are well suited to the bad roads in Sweden. The
medical arrangements and appliances in these ambulances are
thoroughly up to date.

Medical and transport units are united in Sweden, and form
altogether six corps (*trängkärer*), one for each division. Each
corps consists of two transport companies and one medical
company.

A medical company is composed of (*a*) stretcher bearers, and
(*b*) dressing station. (*a*) is divided into two ½ companies of two
sections each. The sections are divided into three groups and
each group forms two patrols. This organization corresponds
to that of the infantry of a division, viz. :—

> One ½ company for each brigade.
> One section for each regiment.
> One group for each battalion.

The group consists of two patrol leaders, and 16 stretcher
bearers with eight stretchers.

Health of the troops.—The number of sick was comparatively
small. Out of 38,000 men engaged, there were 70 sick of whom
only two were serious cases.

General Impressions.

The higher commanders in Sweden have but few opportunities
of handling large bodies of troops, and want of experience in
this respect was noticeable.

The officers of all arms produced an excellent impression, particularly in the matter of appearance and physique. Though not so thoroughly trained as German officers, their efficiency is of a high order. They are gentlemen, and appear to resemble British officers more closely than do those of other continental armies. In their attitude towards sport, the Swedish officer resembles his British comrade, but the army as a whole is entirely under the shadow and influence of Germany.

Discipline is exemplary, and an excellent feeling exists between officers and men. The following service custom is typical: When a battery commander comes on parade in the morning and receives the report of the senior subaltern, he calls out "Good morning, comrades." Thereupon the whole battery, sitting at attention, shouts out "Good morning, captain." The rule that no alcoholic liquor was to be drunk during the manœuvres was strictly followed, from the King downwards.

The general impression received was that a high state of efficiency prevails in the Swedish Army. Everything was done in a thoroughly business-like way, and there was an air of realism that is often lacking at manœuvres.

SWITZERLAND.

The manœuvres took place from the 2nd to the 5th September in the neighbourhood of Wil, between Zurich and Lake Constance.

The opposing forces, flank divisions of imaginary Red and Blue armies, each comprised 26 infantry battalions, 10 squadrons, 1 machine gun company (pack transport) of 8 machine guns, 12 field batteries, 2 cyclist companies, 2 engineer battalions, 2 telegraph companies, 2 field hospitals, and 2 supply detachments. The total (approximately) of the two forces was 24,000 men, 48 guns, and 5,700 horses.

Operations usually ceased about 1 or 2 p.m., and began again at daybreak, but on one occasion they lasted practically all night.

Organization of staff.—Although the six divisions in the Swiss Army are no longer grouped in army corps, yet the staffs of three army corps are formed in peace, so that in time of war they are available to fill up vacancies in the army staff. This year the 3rd Army Corps furnished the directing staff.

Direction of Manœuvres.

The positions of the imaginary armies were fixed by the Director of Manœuvres each day, and instructions, which were supposed to emanate from the army commanders, were issued by the Director for the two forces.

The total number of umpires allotted to a division was only 9, viz., one to each of the following formations :—Infantry brigade, infantry regiment (3 battalions), artillery regiment (6 batteries) and divisional cavalry (2 squadrons).

Casualties were not practised.

Infantry.

The fire discipline of the infantry appeared good. Orders for range, target, &c., were clearly given. Covering fire was freely used, but several opportunities for it were missed, owing to sections advancing without warning. No rapid fire or snap shooting were seen. All fire was individual, and sights were carefully set.

Attack.—In the attack, rushes were made by sections (of about 40 men) or half-sections, at a rapid pace. At short ranges, about 50-yard rushes were made, but when good cover was near at hand the rush was always made to the cover.

The worst feature of the manœuvres was the way in which attacks were made piecemeal, or with insufficient strength. Three attacks by Blue infantry on the Red position failed, simply owing to lack of numbers. If these three isolated attacks had been combined, at one spot, the position would probably have been captured. The timing of attacks was also faulty. On one occasion two fairly strong attacks were made by Blue, and these attacks used up all the Red local reserves; had they been simultaneous the position might have been captured, but a 15-minute interval between each attack gave Red time to reform and meet the second attack.

Counter-attacks were carried out with great spirit by the local reserves, and, owing to the weakness of the attack in many places, they were most effective. But these counter-attacks were often carried too far, and would in reality have had great difficulty in retreating. In all attacks the men charged with bayonets fixed, officers with drawn swords, and on more than one occasion the opposing sides charged right up to one another; but no accidents were noticed.

The local reserves did not appear to exceed two or three sections* for a front of about 400 yards; this means that the battalion reserve in a battalion section was never more than a company.

The following is an instance of the excellent way in which many of the companies were handled. A company had aided in repulsing an attack, and as soon as the umpire's decision had been given, it was ordered to take up a position at the edge of a wood, facing the retiring troops. The company was distributed by sections; between the sections were placed connecting posts, while the flank men of each section were personally shown the positions of these connecting posts. A patrol was then sent out to follow up the retreating troops; the men were told to get well under cover; and finally the range to a wood, where the enemy might collect for a rush, was given out. This company was absolutely invisible from the enemy's position.

Entrenchments.—In both attack and defence the infantry entrenched. At first, the cover obtained was not good, being merely a small non-bullet-proof mound in front of each man. But as time went on, and opportunity allowed, this was improved, and a continuous trench was dug. During a pause in the fight, pioneer companies came up and improved the cover already dug, making an 18-inch trench, traversed at about 12 yards interval, with a good, bullet-proof parapet, 18 inches high. While the pioneers were thus digging, the infantry took cover, ready to re-occupy the trench when needed. A number of trenches were examined. The majority were for fire kneeling, without head-cover, but exceedingly well placed both for fire effect and conceal-ment. A trench could accommodate a half-section (about 20 men) and was covered with grass, bushes, &c., for better concealment. As the type of trench described would give little cover against distant artillery or infantry fire, it is probably only meant for hilly country, where the attacker is on a much lower level.

Marches.—The march discipline appeared to be good; no strag-glers were seen; troops and wagons kept well on the right of the road, and the column, in spite of the hilly nature of the road, was well closed up. When halted, the infantry at once piled arms and took off their packs.

---

* There are 4 sections (*Zug*) in a Swiss company.

The commanders of the advanced guard and other formations always reported on passing a general or staff officer what their command was and what it was doing.

Colours.—On the march the battalion colour was carried, cased, by an under-officer. In a bayonet attack it was uncased and carried forward with the charging line. It is the Swiss national flag, about three feet square, attached to a light pole and with the name in gold of the regimental district.

Cyclists.—Cyclist companies on both sides were largely used for inter-communication purposes. A number of cyclists were allotted to staff officers and were often left at road junctions to show the road and give information.

Cavalry.

The cavalry were well horsed. Their riding, drill and horse-mastership appeared good. The tactical work performed during the manœuvres did not appear good, and the senior leaders did not exhibit much initiative. The Swiss officers recognized this and hope for improvement,

The cavalry private owns his horse, and looks after it throughout the year.

Artillery.

The field batteries were composed of 4 guns and 10 wagons. They appeared to be as well organized and trained as in many regular armies. Over difficult ground the drivers rode well, without fuss or excessive use of the whips; the traces were kept taut, and all the horses in the collar. Teams were even and well up to their work, although the artillery horses are hired and have only a few days' training as such prior to manœuvres.

Semi-covered positions were the general rule, but batteries were manhandled forward into full view whenever necessary.

The tactics resembled those of the Germans very closely, and guns were brought out into the open to fire at infantry within as close range as 400 yards.

A Swiss officer stated that all artillery orders were issued by the Divisional Commander, after obtaining the advice of his senior artillery officer.

No artillery patrols were seen and there are no observation ladders (except with heavy artillery).

On manœuvres, ammunition wagons and limbers do not contain any service ammunition.

Engineers.

The engineer units laid cables (*see* below), supplied tools for entrenchments and supplemented the work of the troops in construction. No wire entanglements were seen.

Signal Service.

A telephone line was laid beside the road from front to rear of columns, the cable being paid out from rolls on the men's backs. Apart from this, inter-communication appeared to have been limited to mounted orderlies and cyclists. There was no visual signalling.

Supply and Transport.

During the manœuvres the troops were always within a few miles of a railway, and the regimental wagons and travelling kitchens marched every day to the railway to receive supplies.

The troops can cook for themselves in their mess tins if necessary. Usually, however, the food is prepared in the travelling kitchen (the body of a limbered vehicle), which cooks for 200 men, and has two boilers, one generally used for coffee, the other for soups and stews. No baking, frying or grilling is possible, but the men do not want it, and the best value is obtained from the food by soups and stews.

The troops generally get coffee, and perhaps some bread or soup, in the early morning. The kitchens then go off for rations, which they cook on the road back. The principal meal is then ready and can be issued when required.

Medical.

Regulations exist for the construction of latrines and burning of refuse, but during manœuvres, when troops were not longer than one night in a locality, nothing apparently was done.

No arrangements for sterilizing or filtering water were seen. Officers said none existed.

General Impressions.

The system of organization in Switzerland on the militia basis is well known and requires no further explanation. Opinions differ as to the value, for continental warfare, of the troops produced, but for purposes of home defence and maintenance of neutrality, the measures adopted probably are sufficient. Mobilization is rapid; in the present instance, the infantry and cavalry assembled at their mobilization stations on the morning of the 24th August and were ready to move off in the early afternoon. For the artillery and transport, whose horses are hired, a longer period—two days—was required.

The Swiss army has very few regular officers, but the system which requires extra training for every step of rank has produced good results. An officer is required to make very considerable sacrifices, but no attempt is made to render the rank an ornamental one by decreasing the qualifications.

The operations were quiet, businesslike and thorough. For 14 days the troops, many of whom were factory hands, were in the open; drilling or manœuvring most of the day, and sometimes part of the night; sleeping in the open or in barns, with rain about half the time and the long grass or crops always wet. Yet the work was done cheerfully and well.

When questioned, the Swiss shows no special enthusiasm about military service; in many cases he will tell you of the sacrifices it involves; but if you ask him whether he would vote for its abolition, he will reply emphatically—"No." Not love of soldiering, but a determination to maintain the independence of his country is what makes the men of Switzerland do their service almost as an act of religion.

The result of all this work, though it does not attain a surprisingly high standard, produces an army whose average age is over 26 years, and minimum standard of training is probably better than that of any large body of non-regular troops in the world.

UNITED STATES.

The manœuvres took place in Connecticut between the 10th and 19th August.

About 20,000 men (divided into two forces) were engaged; they belonged to the Organized Militia or National Guard, with the exception of the following regular troops:—1 cavalry regiment, 1 infantry regiment, 1 field artillery battalion, and detachments of engineers, signal corps and sanitary troops.

Movements of troops were controlled by the directing staff during the first four days; thereafter divisional commanders assumed entire control of their commands.

As the majority of the troops belonged to the National Guard, marches were curtailed and the conditions of field service rendered as light as possible. Movements of main bodies were not permitted on any day before 7 a.m.

Direction of Manœuvres.

Umpires.—Regular officers who performed the double rôle of umpires and instructors were attached to all units of the National Guard not commanded by regular officers. They were specially directed to see that officers and men were fully informed each evening of the events of the day, and that the lessons to be learnt were pointed out. A narrative of each day's operations was given to each officer.

Manœuvre regulations.—The acquisition of information from civilians was forbidden.

Motor-cars were not allowed in front of advanced guards or outposts. The use of motor-cycles was restricted to messengers.

Prisoners were retained until the end of the day's operations.

War diaries were kept by regiments, and by separate organizations throughout the manœuvres.

Infantry.

Regulars.—The advance to attack of the regular infantry over rough ground was methodically carried out in successive rushes from cover to cover, and supported by fire when possible. After each rush the men were marshalled under cover; the supports followed close behind the firing line, and communication between the two was maintained by orderlies.

The general fire control seemed good. Clear and definite orders as to targets were given, but there was no special system of designating them.

No visual signalling was used.

National Guard.—No particular stress appeared to be laid on the seizure of tactical features in order to assist infantry in the attack. Obvious points of advantage were ignored, even when

not held by the enemy; and when they were so held, no real
efforts were made to secure them, though it must have been
evident that the attack could not proceed without unnecessary
loss unless they were at least contained.

The men were generally extended at varying intervals in long
lines, and often without supports. When supports were furnished,
co-operation between them and the firing line was not apparent.
Firing began as soon as the lines were extended. Only on one
occasion was any serious attempt made to build up a strong firing
line within 600 yards of the enemy's position, assisted by a strong
covering fire (at a range of 800 yards) from troops under cover.
There was no methodical advance from cover to cover, or from
one firing position to another; and, generally speaking, no
intelligent use was made of the ground except by individuals.
The control of fire was indifferent, and the amount of volley firing
was very noticeable.

Outposts.—The outpost system was the same by day and
night. Main approaches were strongly held, but the intervening
country was not watched, even by day. Outposts were thrown
well out from the bivouacs and strongly supported, but were not
entrenched. Mounted patrols from infantry regiments were frequently sent out considerable distances.

Cavalry.

The country was not suitable for cavalry, and this arm was
usually employed on wide turning movements for the purpose of
attacking the hostile flank and rear. There was no thought of
close co-operation with other arms, and the cavalry acted
independently in the most literal sense.

. Such duties as advanced, flank and rear guards generally were
performed in an academic manner, and the drill book distances were
maintained without regard to tactical requirements. Scouting
and patrol duties between the two forces were carried out
intelligently, but as the distance was so small, no real opportunities for reconnaissance were afforded.

The horses carried an average weight of 265 lbs., which does
not include the cloak (13 lbs.). The system of horse management was not good; little grooming was done, and feeds were
placed on the ground, or in nosebags, as suited the convenience of
the rider; rope galls were numerous. No watering troughs were
provided.

Artillery.

Direct fire was the rule. On the few occasions when indirect
fire was used, the guns were so near the crest that their flashes
were easily visible, and the positions quickly located by the enemy.
Observation ladders were not used, and there was no attempt to
entrench. Batteries communicated with the brigade commander
only by telephone. There was no system of communication
between artillery and infantry.

Visual signalling was not employed; it was considered slow
and unreliable. The same opinion seemed prevalent as regards
telephonic communication.

The armament and equipment of the artillery of the National Guard were similar to those of the regulars, but the horses were far inferior, a large proportion being hired. By the end of manœuvres they were jaded wrecks, and would not have lasted a week on active service.

Air Service.

Aeroplanes were used for the first time at manœuvres. The country was, however, thickly wooded and suitable landing places were scarce. No passengers were carried.

The machines used were—

Two Curtis biplanes (one privately owned), 75-h.p. engines.
One Burgess-Wright biplane, 35-h.p. engine.

An experimental wireless apparatus was attached to the Burgess-Wright biplane but little was done with it, except to communicate with the artillery with regard to the observation of fire.

The following appear to be the chief lessons gained as regards aeroplanes :—

Aeroplanes cannot supplant cavalry in scouting and reconnaissance work at present, owing to their dependence on atmospheric and weather conditions.

Great stress is laid on the necessity of good climbing capacity, which is expected to be the deciding factor in aerial combat.

Trained military observers are essential.

Biplanes are considered preferable to monoplanes; only the former are used by the Signal Corps.

A clear space for alighting is required.

Powerful brakes are necessary in order to stop quickly in bad landing places.

Each aeroplane should have a wireless equipment.

Closed bodies to protect the operator and the observer are recommended.

Large canvas hangars are unnecessary except in permanent camps; on other occasions, machines can be staked down, and engines covered with tarpaulins.

Each aeroplane requires a motor lorry (60 h.p.) to transport fuel, spare parts, &c.; and one motor for every two aeroplanes is required to convey the personnel necessary for reconnoitring landing places, transmitting information, &c.

Each aeroplane requires a crew of five, viz.: 1 serjeant, 1 corporal, and 3 privates. Each group of three or four aeroplanes should have one supply officer and one or two supply serjeants.

Supply and Transport.

A supply depot was formed for each force. Supply trains were made up at each of these depots, and dispatched to stations near the troops, whence wagon columns conveyed the supplies to camp.

At each depot there was a field bakery capable of turning out 8,000 4-lb. loaves of excellent bread daily, and manned by bakers trained at the army school. This bread has a thick crust and in consequence remained fresh for 10 days or more ; it is much liked.

Frozen meat was extensively used.

There was no organized system of mechanical transport; some motor lorries were hired, but the remainder of the transport work was carried out by the ordinary service wagons, supplemented in some cases by hired carts.

Each unit moved with its own wagons, and, except in the case of regulars, contracted for its own animals. The drivers were quite unskilled, and were the chief cause of the failure of the transport arrangements. There was no regimental supervision, and wagons were often much overloaded.

The body of the service wagon is detachable from its wheels, and with the aid of a tarpaulin can be made into a punt.

One regiment had a travelling kitchen which was a great success.

Medical.

The field hospitals and ambulance companies carried out their work on lines similar to those in force in the British service. Casualties in action were practised, as was the dressing and removal of the wounded.

Sanitation in camp was not of a high order, and latrine accommodation often was lacking.

No water-carts were available, and water, after inspection by a medical officer, was brought from the nearest source in buckets. There was no sterilizing apparatus.

INDEX.

A.

F.

Feet—   PAGE
  care of 84
  of reservists... 65
Field artillery (*see* Artillery).
Field glasses—
  with artillery units 76, 109
  with machine gun detachment 106
Filters, none 84, 131
Flank guard, composition 17
Forage—
  cord carried by cavalry 23
  ration 59
Fortress manœuvres 3
Front, extent of 7, 17, 18, 70

G.

General advanced guards 15
General reserve, use of 16, 100
Grocery wagon 5
Guide officers 113

H.

Hand grenades 103
Hay—
  bales easily divisible 82
  supply of 37, 111
Headquarters (*see* under Staff).
Heavy artillery 28, 94
Horses—
  accommodation in villages... 46
  boarded-out 93, 95
  heel-pegs not used 23, 120
  method of picketing 75
  name painted on saddle 23
  owned by cavalry privates 125
  weight carried by 93, 107, 116, 129
Horsemastership... ... ... 22, 75, 120, 129
Howitzers, use against aircraft 108

I.

Imaginary formations 51
Infantry—
  ammunition 58, 103, 115
  attack 4, 53, 70, 101
  characteristics of 4, 53, 58, 115
  counter-attacks 124
  covering fire, neglect of 70, 102
  cycle orderlies 71
  defence 53
  extensions in attack 102, 115
  entrenchments 20, 70, 124
  equipment 58, 92, 101
  equipment, weight of 58, 101
  fighting in woods 92
  fire in attack 103
  fire positions in attack 102, 129
  machine guns 20, 53, 58, 71, 93, 104, 116
  marching 19, 71, 124
  mounted scouts with 20, 129